T0146345

Federal Financial Incentives to Induce Early Experience Producing Unconventional Liquid Fuels

Frank Camm, James T. Bartis, Charles J. Bushman

Prepared for the United States Air Force and the
National Energy Technology Laboratory of the
United States Department of Energy

RAND PROJECT AIR FORCE and
INFRASTRUCTURE, SAFETY, AND ENVIRONMENT

The research described in this report was sponsored by the United States Air Force under Contract FA7014-06-C-0001. Further information may be obtained from the Strategic Planning Division, Directorate of Plans, Hq USAF. It was also supported by the National Energy Technology Laboratory, United States Department of Energy, and was conducted under the auspices of the Environment, Energy, and Economic Development Program within RAND Infrastructure, Safety, and Environment.

Library of Congress Cataloging-in-Publication Data is available for this publication.

ISBN: 978-0-8330-4510-2

The RAND Corporation is a nonprofit research organization providing objective analysis and effective solutions that address the challenges facing the public and private sectors around the world. RAND's publications do not necessarily reflect the opinions of its research clients and sponsors.

RAND® is a registered trademark.

Published 2008 by the RAND Corporation
1776 Main Street, P.O. Box 2138, Santa Monica, CA 90407-2138
1200 South Hayes Street, Arlington, VA 22202-5050
4570 Fifth Avenue, Suite 600, Pittsburgh, PA 15213-2665
RAND URL: http://www.rand.org/
To order RAND documents or to obtain additional information, contact
Distribution Services: Telephone: (310) 451-7002;
Fax: (310) 451-6915; Email: order@rand.org

Preface

The U.S. Congress and federal agencies are considering a variety of legislative proposals to promote the development of unconventional fuels in the United States. To inform the public discussion of these proposals, the U.S. Air Force and U.S. Department of Energy asked the RAND Corporation to examine the issues and options associated with establishing a commercial coal-to-liquids (CTL) industry within the United States. This technical report is part of a broader effort, documented in Bartis, Camm, and Ortiz (2008), which describes the technical status, costs, and performance of methods that are available for producing liquids from coal; the key energy and environmental policy issues associated with CTL development; the impediments to early commercial experience; and the efficacy of alternative federal incentives in promoting early commercial experience. In support of that companion book, this technical report provides additional detail on the design and assessment of federal financial incentive packages that can successfully promote early commercial experience with CTL production.

The research reported here was sponsored by the Deputy Chief of Staff for Logistics, Installations and Mission Support, Headquarters, U.S. Air Force, in coordination with the Air Force Research Laboratory, and by the National Energy Technology Laboratory, U.S. Department of Energy. It was conducted within the Resource Management Program of RAND Project AIR FORCE and the RAND Environment, Energy, and Economic Development Program (EEED) within RAND Infrastructure, Safety, and Environment (ISE).

This technical report should interest public-policy analysts and decisionmakers seeking additional detail on the methods used to design and assess alternative financial-incentive packages in Bartis, Camm, and Ortiz (2008). More broadly, it should interest analysts seeking more-robust ways to design and assess public policies in an environment with high uncertainty about future resource costs and prices, environmental pressures, and technology cost and performance, all of which can affect the outcomes of any public policy.

This report builds on earlier RAND Corporation publications on natural resources and energy development in the United States. Most relevant are the following:

- *Oil Shale Development in the United States: Prospects and Policy Issues* (Bartis, LaTourrette, et al., 2005)
- *Understanding Cost Growth and Performance Shortfalls in Pioneer Process Plants* (Merrow, Phillips, and Myers, 1981)
- *New Forces at Work in Mining: Industry Views of Critical Technologies* (Peterson, LaTourrette, and Bartis, 2001)
- *Producing Liquid Fuels from Coal: Prospects and Policy Issues* (Bartis, Camm, and Ortiz, 2008).

RAND Project AIR FORCE

RAND Project AIR FORCE (PAF), a division of the RAND Corporation, is the U.S. Air Force's federally funded research and development center for studies and analyses. PAF provides the Air Force with independent analyses of policy alternatives affecting the development, employment, combat readiness, and support of current and future aerospace forces. Research is conducted in four programs: Force Modernization and Employment; Manpower, Personnel, and Training; Resource Management; and Strategy and Doctrine.

Additional information about PAF is available on our Web site:
http://www.rand.org/paf/

The RAND Environment, Energy, and Economic Development Program

The mission of RAND Infrastructure, Safety, and Environment is to improve the development, operation, use, and protection of society's essential physical assets and natural resources and to enhance the related social assets of safety and security of individuals in transit and in their workplaces and communities. The EEED research portfolio addresses environmental quality and regulation, energy resources and systems, water resources and systems, climate, natural hazards and disasters, and economic development—both domestically and internationally. EEED research is conducted for government, foundations, and the private sector.

Information about EEED is available online (http://www.rand.org/ise/environ).

Questions or comments about this book should be sent to the project leader, Frank Camm (Frank_Camm@rand.org).

Contents

Figures

Tables

Summary

This technical report explains an analytic way to design and assess packages of financial incentives that the government can use to cost-effectively promote early experience with coal-to-liquids (CTL) production of liquid fuels in the face of significant uncertainty about the future. It provides technical support to Bartis, Camm, and Ortiz (2008), which places early CTL production experience in a broader policy context.

Analytic Methods

The report applies two complementary analytic methods. The first uses observations from successful voluntary agreements in the commercial world to identify principles that the government can use to design a relationship with a private investor that is likely to ensure that early CTL production experience occurs cost-effectively. Such a relationship yields investor and government behavior that, in turn, generates a set of cash flows to and from investor and government over time. The second analytic method takes these cash flows as given and assesses their effects on the investor and the government. It measures effects on an investor in terms of changes in the investor's real (adjusted for inflation) after-tax internal rate of return (IRR). It measures effects on the government in terms of changes in the real net present value (NPV) of cash flows to and from the government when assessed at the discount rate set by the Office of Management and Budget (OMB) for investments of this kind.

Principles for Designing Incentives

The principles identified in the first half of the analysis and their implications for public policy are the following (see pp. 3–5):

- The more control a specific party to an agreement has over a particular risk, the greater responsibility that party should have to mitigate that risk. For us, all else equal, the more control an investor has over the design, construction, and operation of a CTL production plant, the more the investor should benefit from success or pay for failure in each of these phases.
- The more risk averse a specific party to an agreement is relative to other parties, the more the agreement should shift risk from the risk-averse party to others. For us, all else equal, because the government will typically be less risk averse than an investor will, public policies should seek opportunities to shift risk to the government.

- An agreement should seek opportunities to limit the cost of managing the agreement itself. For us, all else equal, the government should seek to use existing government structures and organizations that implement incentives (such as the tax code and Internal Revenue Service) instead of designing incentives that will require new government structures and organizations.

- Where one party to an agreement has some cost advantage over the others, the agreement should seek to exploit that advantage. For us, all else equal, because OMB prescribes a government discount rate that is likely to be lower than an investor's costs of capital are, the government should seek opportunities to help an investor as early as possible in a project, potentially in exchange for rewards to the government later in the project.

- Parties with a larger stake in an agreement should give special attention to the performance of those with a smaller stake. For us, all else equal, the government should increase its oversight, in source selection and project execution, as an investor uses more debt to finance the project and increase it still further if the government offers a loan guarantee.

- An agreement should seek to adjust to external changes in ways that encourage all parties to remain in the agreement as long as adjustments can be made that allow all to continue benefiting from it. For us, all else equal, the government should not design policies that could force an investor to repeatedly lose money during operations or allow an investor to receive, through a government program, what could easily be perceived to be excessive profits over a long period.

Taken one at a time, these principles often point in different directions. The best policy design seeks to apply these six principles in a balanced way.

Analysis of Cash-Flow Effects in Alternative Futures

The cash-flow analysis focuses on a hypothetical CTL combined-cycle production plant that uses a Fischer-Tropsch (FT) technology to convert coal into about 30,000 barrels per day (bpd) of diesel and naphtha; significant amounts of electricity, some of which can be sold off site; and carbon dioxide, which can be sequestered or sold for use in enhanced oil recovery (EOR) off site. We take the engineering details on the plant from a recent Southern States Energy Board (SSEB) report and add our own assumptions about construction and operational costs; project financing; tax treatment; future prices for coal, oil, electricity, and carbon dioxide; and so on. The analysis yields a set of cash flows over five years of plant construction and 30 years of plant operation that we can use to assess effects on an investor and the government. Because significant uncertainties exist, the analysis considers these effects across a broad range of potential values for real average oil prices and carbon dioxide costs and project costs over the life of the project.

Using this cash-flow analysis, we seek packages of financial incentives with the following characteristics: They increase returns to investors in futures in which cash flows would not induce an investor to pursue early CTL production experience. They limit public-policy effects on investors in futures in which cash flows are likely to induce anyone to invest without government intervention. They seek to emulate a kind of insurance policy in which (1) the government pays companies to invest if private cash flows alone are not sufficient to induce private investment and, in return, (2) companies pay the government a share of their profits when private cash flows alone do induce private investment. Packages with these characteristics allow the government to achieve its primary goal while limiting the expected taxpayer cost of doing so.

The analysis allows us to assemble financial-incentive packages from the following policy components and compare their joint financial effects on investors and the government in different futures:

- a purchase guarantee with a preset purchase quantity and fixed price for the CTL fuel
- a price floor with preset purchase quantity for CTL fuel
- various subsidies that reduce the private firm's investment cost
- a subsidy that reduces the private firm's operating cost
- an agreement to share net income, under preset, specified circumstances, between the private firm and the government when oil prices are high
- a government loan guarantee for a preset portion of the private firm's debt financing.

One particular metric proves to be especially helpful in the design and adjustment of incentive packages to meet these goals. For any change in an incentive package, it measures the cost to the government of raising real private after-tax IRR by one percentage point in any future. Using this metric to compare the government's costs of increasing private IRR in different ways facilitates comparing specific incentive-package changes and ultimately allows designing packages that embody these characteristics.

Policy-Relevant Findings and Recommendations

A balanced package of a price floor, investment subsidy, and income-sharing agreement would allow the government to achieve its primary goal of ensuring early CTL production experience at a reasonable cost to the government. (See pp. 43–49.) The investment subsidy is a cost-effective way to raise private after-tax IRR in any future. A price floor can cost-effectively provide an additional boost in futures in which oil is especially inexpensive. And an income-sharing agreement can effectively complement any investment subsidy and price floor to create a kind of insurance agreement between the investor and government. In such an agreement, the government effectively offers an investment subsidy and price floor to insure the investor against loss during years with low prices in exchange for a share of investor profits for years with prices high enough to justify such sharing.

Among investment incentives, those that convey benefits to an investor early are the most cost-effective for the government. (See pp. 27–31.) The cash-flow analysis confirms the expectation just described, that OMB's low government discount rate favors the use of investment subsidies. Investment subsidies, such as tax credits, take advantage of this difference to help investors finance their investments. Accelerated tax depreciation is not nearly as cost-effective. And subsidies tied to production are still less cost-effective than any of the investment subsidies examined. Production subsidies prove useful only if required to maintain investor interest in production after the plant is built. Net revenues from our hypothetical project were high enough across all considered futures to sustain production with no additional government support.

Loan guarantees can strongly encourage private investment. But they encourage investors to pursue early CTL production experience only by shifting real default risk from private lenders to the government. (See pp. 33–42.) By their very nature, the more powerful their effect on private participation in a project, the higher their expected cost to the government. And loan guarantees encourage private investors to seek higher debt shares that, by increasing their default risk, raise the government's expected cost of any loan guarantee. The government should use loan guarantees to promote early CTL production experience only with great care and with a full appreciation of their potential costs to the Treasury and the extent to which government oversight of guaranteed loans effectively limit those costs.

Because the exact form that a balanced package would take depends on expectations about project costs, the government should wait to finalize its design of an incentive package until it has the best information on project costs it can get without actually initiating the project. (See pp. 43–48.) We strongly advise that an incentive agreement not be finalized until both government and investors have the benefit of improved project-cost and performance information that is provided after a front-end engineering design (FEED).

Some investors will be significantly more likely to achieve early CTL production experience than will others. For example, we expect more technologically sophisticated investors with more experience building and operating first-of-a-kind chemical plants and that have a long-term stake in exploiting the knowledge gained from early CTL production experience to be more likely to succeed than investors looking primarily for an investment opportunity that fits well in a broader financial portfolio. (See p. 14.) They would certainly be more likely to succeed than small and disadvantaged businesses in general.

The government should clearly pursue a preference for investors that are more likely to achieve its primary goal—early CTL production experience—in the acquisition strategy it builds for choosing investors to support. That strategy should include thorough due diligence regarding the technological, management, and financial capacity of all competitors. It might go further and allow offerors, as part of their proposals in source selection, to design parts of the incentive package the government uses to oversee and reward the chosen investors. (See pp. 51–52.) That is, using the performance-oriented approach that the federal government now prefers in source selection, this strategy would substitute a statement of objectives, which states what the government values in a new investment in a CTL plant, for a statement of work, which specifies how the government would reward the winner of such a source selection. Properly designed and implemented, such an approach to source selection could give the government valuable insights about each potential investor's priorities, beliefs, and capabilities and help it choose a package of financial incentives best meeting the mutual interests of each investor and the government.

Acknowledgments

We benefited from ongoing, informal interaction with Keith Crane and Michael Toman throughout the project. Discussions related to another risk-related project—on the application of formal risk-assessment methods to U.S. Air Force strategic force planning—provided serendipitous insights on this one. Useful discussions occurred with Lauren Casten, Alexander C. Hou, Forrest Morgan, and Alan J. Vick throughout this project. More structured, extended meetings with Paul K. Davis, James A. Dewar, Lloyd Dixon, Andrew R. Hoehn, Tom LaTourrette, James T. Quinlivan, and Robert R. Reville were also helpful. Discussions with Paul Davis were invaluable to the approach we ultimately chose to represent the effects of uncertainty on policy outcomes. Technical reviews by Keith Crane, James Ekmann, and Hillard Huntington have materially improved the report.

We thank them all and retain full responsibility for the accuracy, analytic soundness, and objectivity of the work reported here.

Abbreviations

bpd	barrels per day
CO_2	carbon dioxide
CNR	Canadian Northern Railway
CRIOP	cost of raising private, after-tax IRR one percentage point
CTL	coal to liquids
DDB	double-declining balance
DVE	diesel value equivalent
EEED	RAND Environment, Energy, and Economic Development Program
EIA	Energy Information Administration
EOR	enhanced oil recovery
FEED	front-end engineering design
FT	Fischer-Tropsch
IPCC	Intergovernmental Panel on Climate Change
IRR	internal rate of return
ISE	RAND Infrastructure, Safety, and Environment
kWh	kilowatt-hour
MACRS	Modified Accelerated Cost Recovery System
MW	megawatt
MWh	megawatt-hour
NPV	net present value
OMB	Office of Management and Budget
SSEB	Southern States Energy Board

CHAPTER ONE

Introduction

Rising petroleum prices have prompted interest in using coal to manufacture liquid fuels that can displace petroleum-derived gasoline and diesel fuels. Coal is abundant in the United States and elsewhere, and coal-to-liquids (CTL) technology is commercially viable. But great uncertainties persist about the cost and performance of new CTL production facilities, the price of petroleum over the life of such facilities, the value or cost of carbon dioxide (CO_2) coproduced with liquid fuels in a CTL facility, and other factors relevant to the economic viability of new CTL production facilities.[1] In the face of such uncertainty, this technical report describes a way to design financial-incentive packages that could ensure project viability at limited cost to the government. In particular, it provides technical details that underlie the discussion of policy design in Bartis, Camm, and Ortiz (2008).

The Intergovernmental Panel on Climate Change (IPCC) noted,

> Financial incentives are frequently used by governments to stimulate the diffusion of new . . . technologies. While economic costs are generally higher for these than for other instruments, financial incentives are often critical to overcoming the barriers to the penetration of new technologies (*high agreement/much evidence*). (Gupta et al., 2007, p. 747. Emphasis in original.)

It noted that, because individual policies rarely operate in complete isolation,

> many cases require more than one instrument. For an . . . effective and cost-effective instrument mix to be applied, there must be a good understanding of the . . . interactions between the different instruments in the mix. Applicability . . . can vary greatly, but may be enhanced when instruments are adapted to local circumstances (*high agreement/much evidence*). (Gupta et al., 2007, p. 748. Emphasis in original.)

This report focuses on packages comprising the following financial-incentive instruments:

- a purchase guarantee with a preset purchase quantity and, potentially, a fixed price for the fuel
- a price floor with preset purchase quantity for the fuel
- various subsidy programs that reduce the private firm's investment cost
- subsidy programs that reduce the private firm's operating cost

---

[1] For several recent perspectives on how to induce investment in unconventional technologies in the face of uncertainty, see Blyth and Yang, 2006; Hamilton, 2005; and Reedman, Graham, and Coombes, 2006.

- an agreement to share net income, under preset, specified circumstances, between the private firm and the government[2]
- a government loan guarantee for a preset portion of the private firm's debt financing.

The discussion anticipates that the government will likely use a package of such financial instruments to encourage early CTL production experience.[3] It gives careful attention to how such instruments might work together as a package and how they can be tailored to reflect the specific circumstances relevant to a specific investment.

The report describes qualitative and quantitative factors relevant to designing a package of public policies that would ensure that one or more private investors build and operate unconventional-fuel production plants. It examines how particular financial policy elements and simple packages could affect (1) the real after-tax internal rate of return (IRR) that a private investor could expect and (2) the government's view of the real net present value (NPV) cash flows to and from the government.

Chapter Two discusses qualitative factors that the government can use to help design a package of policy instruments that will sustain a long-term relationship between the government and an investor. This discussion draws on the economic theory of contracting to identify first principles that government policymakers can use to compare incentive packages.

Chapter Three describes the structure of a numerical cash-flow model of an investment in a hypothetical CTL production plant. The model shows how different combinations of financial policy instruments affect a private firm's real after-tax IRR from building and operating such a CTL production facility and the NPV of government cash flows associated with any combination of policy instruments. Appendix A describes this model.

Chapter Four then applies this model to the case in which the investor uses 100-percent equity financing. Chapter Five extends the analysis to circumstances in which the investor uses a mix of debt and equity financing. Appendix B presents mathematical models that present several arguments from that chapter in more formal terms.

Chapter Six draws on the findings in earlier chapters to design two sets of robust financial-incentive packages—packages that reduce uncertainty about outcomes for a private investor. One set assumes low project costs; the second assumes higher project costs. Differences between the packages' designs and performance levels illustrate the importance of gaining good information on project costs before designing an incentive package.

Chapter Seven proposes a way to use a source-selection mechanism that the government might use to design incentive packages. Even if the government declines such a novel approach, thinking about design in the context of source-selection considerations helps clarify the importance of choosing the right investor for a project and tailoring the incentive package that the government offers that investor to the investor's priorities.

Chapter Eight closes the report with a summary of policy-relevant findings.

---

[2]   In our analysis, income sharing gives the government an increasing share of the profit from a plant as the average price of oil rises above a threshold level. The arrangement is analogous to the pricing terms found in many oil-production contracts outside the United States.

[3]   For a useful overview of how such instruments have performed, see Arimura, Hibiki, and Johnstone, 2005. See also Alic, Mowery, and Rubin, 2003.

CHAPTER TWO

Designing an Effective Long-Term Public-Private Relationship

When the federal government seeks to encourage a private firm to build and operate a plant, it faces a "principal-agent" problem; the government wants to induce a private investor to do something in the government's interest. As a principal, the government seeks to design a cost-effective package of financial policies that will induce a private firm—an agent—to build and operate the plant.[1] We take the goal of inducing early CTL production experience as given. The government wants to induce such early experience to kick-start the development of a new industry by accelerating the construction and operation of first-of-a-kind plants. Constructing and operating these plants should accelerate the development of skills, supplier industries, and equipment manufacturing relevant to these industries. We do not question the value of this undertaking; rather, we focus on identifying financial public policies that will (1) induce a private firm to act and (2) limit the government's cost for that induction.[2] In effect, we seek a cost-effective package of policies that will induce a private firm to build and operate the plant in a way that yields the early CTL production experience that the government wants.

First Principles of Incentive-Package Design

The economic theory of contracts can help us anticipate what kinds of public policies would be most cost-effective. This theory seeks to explain the design of voluntary agreements between specific buyers and sellers that survive over time in competitive markets. Presumably, only cost-effective agreements survive. Otherwise, parties choosing voluntarily to do business with one another would choose alternative arrangements from which, by definition, they could all benefit. Available empirical evidence suggests that such agreements tend to have the following characteristics.[3]

---

[1] For an exceptionally well-written, succinct discussion of principal-agent issues, see Dixit, 2002. For a more formal treatment, see Laffont and Tirole, 1993.

[2] IPCC highlights "four main criteria . . . widely used by policymakers to select and evaluate policies: environmental effectiveness, cost-effectiveness, distributional effects (including equity) and institutional feasibility" (Gupta et al., 2007, p. 747). This analysis focuses on cost-effectiveness and distributional effects. Institutional feasibility enters in places, but we address it in terms of cost-effectiveness. Our distributional analysis distinguishes net benefits that accrue to a private investor and to the government. We explicitly avoid combining these net benefits and leave to policymakers the decision about how much government revenue to exchange for an increase in private investor profits in order to promote early commercial CTL development and production.

[3] An excellent survey of the empirical literature on the design on contracts can be found in Masten, 2000. Masten did not explicitly trace out the specific factors highlighted here, but they are consistent with the empirical findings that he reported.

Relative Control

The more control any party to the agreement has over the execution of the agreement, the more responsibility—and therefore risk—the agreement assigns to that party. This approach, in effect, makes the party that is most able to ensure the agreement's success most responsible to ensure that success. Put another way, this principle seeks to have each party bear as large a share of the consequences of its actions—good or bad—as possible. Doing this limits the potential for moral hazard in a relationship. *Moral hazard* occurs in a relationship when one party's pursuit of its own interests injures another party. In our setting, this implies that the more control the investor has over project design and execution, all else equal, the more responsibility and risk should shift to the investor.

Relative Risk Aversion

The more risk averse any specific party to an agreement is, the less risk the agreement assigns to the party. For a variety of reasons, the government is likely to be less risk averse than any private firm that might build and operate the plant. The economy, which frames the government's perspective, is larger and presents a broader and more diversified portfolio than any private investor's portfolio. The economy does not face the threat of bankruptcy in the same way that any individual investor does. And official Office of Management and Budget (OMB) policy reflects this perspective, making it likely that the federal government is less risk averse than are relevant private investors.[4] But large private firms should be better able to bear risk than smaller firms are, because failure of the plant as an investment is less likely to threaten their survival in the marketplace. So, in our setting, all else equal, an incentive package should shift more risk to the government than to the investor. And the smaller the investor drawn to the project, the more risk the government should expect to bear. Insurance offers a common way in which a risk-averse organization can shift risk to another entity better able to bear that risk by aggregating many independent risks to take advantage of the law of large numbers. In our setting, the government could provide such insurance by using an investment incentive or price floor to limit an investor's downside exposure in exchange for receiving a payment from the investor when the private after-tax IRR is high enough to ensure a project's viability even without such payment.

Cost of Relationship

If the parties to an agreement can administer it in ways that reduce its administrative costs without reducing the level of mutual benefits it generates, the agreement should take advantage of such opportunities. In our setting, all else equal, the government should favor policy instruments that are easier to administer—for example, subsidy mechanisms that can be administered through an existing tax infrastructure.

For applications of these principles to practical issues, such as those addressed here, see Goldberg, 1989, and Rubin, 1990. More formal overviews include Bolton and Dewatripont, 2005, and Salanié, 2005.

[4] On relevant government policy, see OMB, 1992. The basis for this policy is explained in Arrow and Lind, 1970. Corporate-finance theorists argue that, in the interests of their shareholders, private firms should be risk neutral, even if doing so invites risk of bankruptcy. Observed organization behavior is rarely consistent with this normative standard, even in very large corporations.

Relative Cost Advantages

If any specific party to an agreement has any special cost advantages relative to other parties, the agreement should take advantage of these when possible. For this type of policy decision, OMB (1992, §8.b.1) seeks to set a government discount rate that "approximates the marginal pretax rate of return on an average investment in the private sector in recent years." Current OMB policy sets that government discount rate at 7 percent in real terms (adjusted for inflation), which is significantly lower than the *pretax* private real cost of capital typically used to assess the value of an investment of the kind considered here. In our setting, all else equal, the larger the difference between public and private discount rates, the more costs the government should accept early in the project relative to the investor.

Relative Size of Stake

This principle addresses the potential for moral hazard from a slightly different perspective. Given any allocation of risk among the parties to an agreement, if any party has a smaller stake in the agreement's success than the others do, the others should plan to assume additional oversight to ensure that success. In our setting, all else equal, given any allocation of risk between the government and the private investor, the smaller the investor's stake in the project, the more due diligence and focused project oversight the government should apply.

Preservation of Relationship

Agreements that can benefit all parties should seek to sustain themselves by encouraging all parties to remain in the agreements. In practice, this principle tends to favor more equitable division of mutual benefit among the parties than can easily be explained by cost-effectiveness concerns alone. It also tends to favor terms that reflect changes outside an agreement (e.g., in prices) that would encourage any party to leave the relationship, even though appropriate changes inside the agreement could allow all parties to continue to benefit from the agreement by staying. In our setting, all else equal, this principle favors terms that respond to unexpected changes in costs, prices, performance, and other external factors, especially changes that encourage the investor to withdraw before enough early CTL experience has accumulated.

Implications for the Use of Alternative Policy Instruments

Taken one at a time, these principles often point in different directions. The best policy design should seek to apply these six principles in a balanced way. As simple as these principles might appear, they provide useful guidance on how to apply the policy instruments we examine here.

Guaranteed Purchases

Suppose the government wanted to guarantee purchases of unconventional oil at some prescribed price. Choosing a price linked to the prevailing market price in the future would ensure that the investor had a ready market for the guaranteed portion of its production. This would not necessarily markedly change the investor's circumstances unless its product were unique in some way and the investor had committed to this plant on the assumption that demand would persist for that unique aspect of its production. A purchase agreement, then, can align the buyer's decisions over the life of the plant (which it can control to some degree) with the seller's decision to build the plant (a decision the builder controlled at the time).

Uniqueness could involve the product's chemical or physical attributes, which the investor might have customized to certain buyer specifications. It could involve location: Perhaps the investor agreed to build in a specific location because it expected demand to continue there. These arguments explain the presence of purchase guarantees in very long-term (40 years and longer) contracts between electricity-generating plants and the coal mines where they are collocated (see, e.g., Joskow, 1987). The generating plants are customized to the attributes of the coal on site; the coal may be worth producing only because of the presence of a collocated generating plant. Similar considerations could apply in the case of an unconventional-fuel plant. In our setting, such an instrument might be most appropriate to government purchase of CO_2. For example, a plant location might be chosen to produce CO_2 that the government could then use nearby in experiments on sequestration.

A purchase guarantee could also specify a price entirely unlinked to market prices for products. For example, a price could be linked to the prices of inputs. This would relieve the investor of risks associated with prices it cannot control. Properly designed, a cost-plus-fixed-fee agreement could focus the investor's attention on the portion of price that it can control and motivate it to optimize its short-term performance against that element of price.[5]

Alternatively, a purchase guarantee could specify a firm fixed price that stood regardless of changes in the prices of inputs or the market price of oil. This increases the power of the incentive the investor faces to react to changes in the prices of inputs it cannot control, inducing the investor to work harder at affecting every element of performance it can control. The incentive that a principal offers an agent is more high powered when it effectively aligns the agent more directly with the principal's core interests, inducing the agent to work harder to promote the principal's interests. Higher-powered incentives induce this effort by exposing an agent more directly to the risks it can mitigate through its own efforts. Even if the agent cannot control the price it pays for an input, it may be able to control how much it uses by changing its production process or product slate.

Such an arrangement can be mutually advantageous when the relevant processes allow effective adjustments and the investor knows these processes better than the government and has more control over their optimal operation. But it can also increase the variance in net income that the investor faces, which may induce the investor to seek higher prices than the government would pay with a cost-based contract. The less certain the investor is about its future costs, the more likely this is to apply. When this occurs, the lower presumed risk aversion of the government suggests that such an increase in price may not be worth the savings created by inducing greater investor control of its assets.

When one organization guarantees to purchase some portion of the production of another, the contract that governs their relationship typically uses a hybrid of these approaches to balance risks between the parties. It typically includes economic price adjustment or cost-escalation clauses for things totally beyond the seller's control, such as the level of local taxes. When a

---

[5]   For example, the investor could use realized *allowable cost* (one certified by auditors to qualify for this arrangement) over a period of time as the basis for setting a firm fixed price that will hold for some set period in the future. Alternatively, the investor could use its expectations about future allowable costs to negotiate a firm fixed price that would hold for some set period of time in the future and submit auditable cost data to justify its estimate of future costs. Contracts described as *cost-plus* contracts often take this form. In either case, once the firm fixed price is set, the investor has significant incentives to optimize its performance against this fixed price.

contract lasts more than a few years, it typically allows adjustment to market prices over the long term to ensure that contract prices do not depart significantly from market prices. Large departures encourage one party or the other to seek a way to terminate and take advantage of opportunities outside; adjustable prices help protect such long-term relationships and the resources both parties have invested in them. They can also help limit the potential for a perception of government-funded excess profits or government-imposed bankruptcy, each with its own concerned constituency. Side by side with this long-term flexibility, such agreements may stabilize prices for a few years at a time to impose discipline on the seller and limit turbulence in the buyer's budgeting process.

As important as these factors are to successful policy design, it would cost too much to collect the data required to reflect them directly in the cash-flow analysis that follows. As policy design goes forward, we should view these arguments side by side with the financial results of the cash-flow analysis and seek an effective synthesis. Neither can prescribe the final policy design alone.

Price Floor

A *price floor* is a variation on a purchase guarantee. With a price floor, the government agrees to purchase some stated quantity with a stated pricing arrangement if the investor chooses to sell at that price. In practice, the investor will sell to the government when the government price exceeds the market price available to the investor and will sell directly to the market when it does not. The pricing arrangement could be a firm fixed price, a cost-plus-fixed-fee price, an economic price adjustment–based price, or some hybrid that adjusts a firm fixed price every few years to reflect longer-term trends in input or output markets.

A price floor is simply a hybrid variation on the purchase guarantees discussed; it gives two parties greater flexibility to split risks between themselves. As a practical matter, it effectively maintains simple links to output-market prices when these prices are high and protects the investor—effectively transfers resources from the government to the investor—when output prices are too low. If the difference in risk aversion is large enough between the government and a specific seller, this can be an especially good way to incentivize the investor at low cost to the government. A risk-averse actor puts greater weight on bad outcomes than on good outcomes. By transferring resources to this investor when prices are low—when the resources are worth more to the investor than to the government at high prices—the government can, to its advantage, effectively exploit a difference between itself and the investor.

The cash-flow analysis that follows demonstrates that a price floor can have large effects on real after-tax investor IRR, often at a reasonable cost to the government. It cannot translate this effect into a measure of how much more important this would be to a relatively risk-averse investor than to a risk-neutral investor.

Investment Incentives

In general, the larger the differences between the discount rates of the government and the specific investor, the more cost-effective investment incentives are likely to be, from the government's point of view,. That is because investment incentives transfer resources from government to investor early in a project. Delaying such a transfer systematically reduces the transfer's value to the investor more than it reduces the government's cost. The cash-flow analysis that follows demonstrates that this factor can be documented as a major consideration in choices among policy instruments.

Investment incentives can come in many forms, and the simple cash-flow analysis we use cannot address important differences among them. We consider them here.

A *lump-sum grant* is a one-time fixed payment that is not conditioned on anything occurring in the project. It is likely to be more cost-effective than is an incentive based on cost sharing, which apportions between government and investor any risk associated with cost changes. That is because, when the investor knows more about its opportunities than the government does, a lump-sum grant generates stronger incentives for the investor to take full advantage of those opportunities to improve plant performance or reduce plant cost. A lump-sum grant allows the investor to capture the full benefit of any innovation. An incentive based on sharing changes in cost forces the investor to share with the government some portion of any benefit from innovation. That said, a cost-based incentive can limit variation in the realized net income the investor experiences. It is clearly preferable to give a risk-neutral investor a lump-sum grant; a cost-based incentive becomes increasingly mutually attractive as the investor becomes more risk averse. The more heavily an incentive relies on information about costs, the more oversight the government has to be prepared to maintain to define and enforce its definition of allowable costs.

Whether an incentive comes as a lump sum or through cost sharing, it can be delivered through a dedicated government program that writes subsidy checks or through the tax system. We will refer to the first version of an incentive as a *direct subsidy* and to the second as a *tax subsidy*. The government makes extensive use of the tax system to do this, because it already exists—no new bureaucracy need be created to administer a new program—and it can make subsidies less visible to the general public. Knowledgeable observers can find any tax subsidy tucked in the tax code, but many subsidies are so artfully written that they will miss the attention of an uninformed eye.

Unless a tax subsidy is codified carefully, use of the tax system can require that the investor have taxable income from outside the project being subsidized, so the investor can use project-generated tax subsidies to offset taxes outside the project. This is more an issue for small investors than for large ones with many sources of taxable income. Current legislation regarding unconventional-fuel production addresses this problem by making certain tax credits salable. Arbitrage ensures that such credits will find their way to taxpayers who value them the most.

When the government uses the tax system to deliver investment incentives, it can choose when to transfer resources to the investor. As noted, the difference in the discount rates of the government and investor favors tax subsidies, such as investment tax credits, that deliver benefits to the taxpayer as early in the process as possible.

Production Incentives

When a specific investor's discount rate exceeds the government's, investment incentives are more cost-effective than are production incentives. As the cash-flow analysis demonstrates, at project start-up, it costs the government substantially less to reduce a project's real after-tax private IRR by one point with an investment incentive than with a production incentive.

But after investment is complete, investment incentives are no longer available. In some projects, a production incentive can help the government ensure that, after investment costs are sunk, an investor still has an incentive to operate the plant it has built. This is the primary role any production incentive is likely to play in an incentive package that promotes private production of unconventional fuels. In a secondary role, an incentive could also be designed to induce more production each year to accelerate the learning process. The government's goals should dictate which form of production incentive to use.

Like investment incentives, production incentives come in many varieties—e.g., lump sum versus cost sharing, direct subsidy versus tax subsidy.[6] Their relative costs and benefits mirror those of investment incentives. As noted, production incentives are most likely to be useful if a plant does not generate taxable net income without government support. As a result, the same concerns raised about the value of tax subsidies to an investor without taxable income arise here. One new wrinkle here is the choice between production incentives rewarding years of production and those rewarding production during any year. The distinction can be important if the incentive package does not effectively dictate, through purchasing and pricing agreements, how much the investor will produce in a year.

A lump-sum subsidy linked to a year is likely to generate the highest-power incentives for the investor. The investor receives the full benefit of any improvement it makes by changing its level of production, production slate, or production methods. It also yields the most information about how a plant might operate without government participation, because lump-sum subsidies are less likely to distort private decisions than are any other types of subsidies. Because the government offers such an incentive to ensure that production occurs, of course, the subsidy must be contingent on some minimum level of production.

If the government actively seeks to accelerate the accumulation of experience at the plant by increasing annual production, a lump-sum subsidy per unit of production creates the highest-power incentives for the investor to do this, for the reasons given already.

As explained, higher-powered incentives tend to increase the level of risk an investor experiences. The more risk averse a specific investor, the more the government will have to pay to take advantage of the benefits offered by high-powered incentives. When the investor's risk aversion is high enough, the higher prices required to induce the investor's participation offset any benefit the government might get from higher-powered incentives.

Net-Income Sharing

Net-income sharing identifies allowable costs and revenues, uses them to calculate net income, and gives the investor and the government each a share of the net income of that year. Such an arrangement is extremely flexible and can be designed in many ways to address the mutual interests of a specific investor and the government.

Technically speaking, under this definition, the federal corporation income-tax system is a net income–sharing arrangement. Typically, a formal net income–sharing arrangement operates alongside the corporation income tax and can use definitions of allowable costs, revenues, and sharing rates entirely different from those in a coexisting corporation income tax. Such arrangements are common in agreements between oil producers and governments outside the United States (Kretzschmar and Kirchner, 2007; see also Metcalf, 2006). These arrangements typically allow the producer to recover some basic costs before any sharing occurs. Then, as higher average oil prices drive a producer's net income higher, the government takes an increas-

---

[6] A lump-sum incentive would pay the investor a fixed sum each year in which the investor produced some threshold amount of liquid fuel. A cost-sharing incentive would measure allowable costs, defined in some specific way, during each year of production and reimburse the investor for some stated share of this allowable cost. A direct subsidy would give the investor a direct cash payment during each year of production. A tax subsidy would instead give the investor some specific tax relief during each year of production.

ing share of the net income that results. For example, a sharing rate might be tied to the producer's real IRR under specific rules about what costs are allowable. In this situation, the government share rises as the real private IRR rises.

Such an approach shares risk between an investor and the government this way: It limits the investor's downside risk by allowing it to use all revenues when oil prices are low so that it can recover operating costs. But as net income becomes available at higher oil prices, it allows both investor and government to benefit from such prices. It seeks to allow a government benefit without ever discouraging the investor from continuing to produce. Caution is required in the use of such agreements, because they can discourage a specific investor from investing in the first place. By reducing the net income the producer would receive if prices were high, such an agreement can reduce the amount the investor would be willing to invest in a production activity, eliminating private-investor interest in some marginal investment.

By definition, we view investment in unconventional-fuel plants as future events whose profitability to a specific private investor depends directly on any income-sharing agreement associated with them. Profits that such an agreement allows directly affect any investor's calculus of how financially attractive the investment might be. As a result, the proper design of any net income–sharing arrangement is of special interest to us. Such an arrangement is most appropriate when coordinated with other policies, such as a price floor, that limit the investor's reliance on the possibility of high prices to justify a new investment. The cash-flow analysis addresses this concern numerically to demonstrate its importance.

The industry has a great deal of experience with such arrangements. That should make it easier for the government to work with experienced producers to frame an arrangement's specific terms—e.g., the definition of allowable costs and revenues, the factors that affect sharing rate—that will promote their mutual interests. Precise definitions are critical to the success of such an arrangement. Fortunately, many effective benchmarks are available to use as starting points.

As net income–sharing arrangements become more complex and affect more parts of a project, at some point, they become essentially joint ventures or public-private partnerships. We will not speculate on when that occurs. We observe only that reducing the arm's-length distance between the government and specific private investors raises important political issues that must be addressed. New public-management efforts to reform federal acquisition policy in the past 20 years have, in effect, encouraged movement in this direction.[7] But such policies remain controversial. Serious abuses, some criminal, have occurred as they have been applied. The government is still learning how to design such arrangements effectively. Knowledge accumulated to date is available to apply to promoting private participation in an unconventional-fuel plant. It should be applied with great care to avoid further complicating an already complicated challenge for reasons irrelevant to the task at hand—getting new plants built to generate early CTL production experience.

Loan Guarantees

If an investor uses only equity capital to finance a project, loan guarantees are irrelevant. If the investor relies on debt capital, however, the government can agree to guarantee the payments on any portion of the loans that that investor plans to use to finance an unconventional-oil

---

[7] For a useful overview of ongoing trends in defense acquisition, see Anderson, 1999.

production project. If the investor can pay back a guaranteed loan, ex post, the guarantee costs the government nothing and the lender achieves its desired rate of return on the loan. If the investor cannot pay back a guaranteed loan, ex post, the government pays some or all of what the lender expected from the investor. Ex ante, the expected cost of such a guarantee to the government is the product of (1) the cost of the loan payments it would have to make if the investor defaulted and (2) the probability that the investor would default.[8]

Such a guarantee may almost fully indemnify the lender, with two consequent effects.[9] First, the lender is willing to offer loans on more favorable terms. For example, the lender might (1) offer the investor a lower interest rate at any level of the investor's debt share of financing or (2) allow a higher debt share at any level of interest rate. The cash-flow analysis examines both types of effects and demonstrates that, in most situations, an investor benefits far more from the second effect than from the first. That is, for the most part, a loan guarantee encourages investor participation by allowing the investor to increase its debt share of financing. This can encourage an investor to undertake a project for which it would not otherwise have the financial resources. Presence of a loan guarantee increases the importance of thorough due diligence to screen firms applying for loan guarantees for their financial, managerial, and technical capacity.

The second effect of near-full indemnification, by so effectively protecting the lender, reduces the lender's stake in the project and its interest in controlling the risks associated with a higher debt share of financing. The higher the debt share, the smaller the stake of the investor or borrower in the outcome. And the more complete the indemnification of the lender, the smaller its stake. Such changes generally violate the principles of assigning risks to the parties of an agreement most able to affect them. They also violate the principle of increasing oversight over parties with small stakes unless the government accompanies any loan guarantee with a significantly expanded oversight role in the project.

In effect, a loan guarantee makes the government an important capital claimant. Unless the government takes this role seriously, use of a loan guarantee can very easily increase a project's risks of failure by not imposing the discipline otherwise provided by the private-sector capital claimants in the project. To avoid this outcome, the government must act to maintain the discipline that the market provided in the absence of loan guarantees.

This role demands a degree of government oversight and involvement in the project that exceeds that for any of the instruments described here, except some of the more complete forms of partnership. Because a loan guarantee can attract specific investors with a smaller long-term stake in the potential up- or downside outcomes of a project and so with fewer skills relevant to the project's success, the government must perform more complete oversight and due diligence in any process used to select investors to build plants; this issue of *adverse selection* is well known in the literature on designing incentive systems.[10] Once the government has selected a specific investor, the government must maintain more complete oversight to ensure that the investor performs in a way that promotes the government's interests and not, given the attenu-

---

[8] Because we assume the government to be risk neutral, the expected value of loss to the government is all we need to know about this. The probability of default, of course, is a highly subjective component of any such calculation.

[9] Even if the federal government agrees to pay anything the investor cannot, the administrative process and delays associated with it will inevitably hurt the lender. So even a full government guarantee will not fully indemnify the lender.

[10] Dixit, 2002, provided a good discussion of this point.

ated nature of the investor's role as a capital claimant, the investor's interests—that is, to limit the potential for moral hazard.

In effect, a loan guarantee makes the government a partner. Such involvement is costly, in terms of both resources and senior government-leadership focus. We cannot quantify such costs in any convincing manner that would allow us to include them in a cash-flow analysis. But we can subjectively weigh their importance against any potential benefit associated with a loan guarantee when only strict financial considerations are examined. If the government fails to perform as an effective partner, a loan guarantee imposes another kind of cost on the government by increasing the possibility of failure. A cash-flow analysis can capture the size of this cost in terms of the government's obligations to repay any loan; the cash-flow analysis numerically illustrates the nature of this cost. Such an analysis cannot capture the cost associated with a project failure severe enough to threaten the successful construction and operation of the plant. If the plant is not constructed and operating, the project cannot generate the early CTL production experience that motivated interest in such a plant in the first place. The cash-flow analysis we use does not capture the value of such experience.

CHAPTER THREE

Assessing Financial Effects Under Uncertainty

The qualitative factors described in Chapter Two can help us choose what type of financial instruments to consider in any incentive package. Detailed cash-flow analysis allows us to assess the effects of choosing specific values for the attributes of these instruments—e.g., the level of a price support, the number of barrels in a purchase guarantee, the size of a tax credit, the specific terms of a net income–sharing agreement. As we go forward, please keep in mind that the cash-flow analyses presented here do not attempt to capture the negative effects of moral hazard and adverse selection discussed in Chapter Two. In effect, the analysis in this chapter accepts some level of both and assumes that whatever is present is fully captured in our assumptions about flows of relevant revenues and costs. But it does not attempt to adjust these flows to reflect the potential for, for example, rising operating costs through the effects of moral hazard as the government share of operating costs rises, or lower effective plant availability as the private investor's stake in the project falls. A full appreciation of the effects of alternative incentive packages depends on an integrated application of the quantitative and qualitative methods presented here.

Basic Design of the Cash-Flow Analysis

The Basic Project

The analysis focuses on investment in and operating of a specific type of combined-cycle plant, described in detail in a Southern States Energy Board (SSEB) report.[1] The plant gasifies coal, uses the Fischer-Tropsch (FT) method to convert the gas to liquid fuels, and produces electricity for use on site and export. It has the following characteristics:

- daily production of 30,000 diesel value equivalent (DVE) barrels of diesel and naphtha; 725 megawatt-hours (MWh) of power, 204 of which are exported; and 24,734 tons of CO_2 not consumed in the process
- daily consumption of 17,987 tons of bituminous coal.

The performance of the first-of-a-kind CTL that we examine is consistent with that of the plant that the SSEB described in its report. But we adjust its cost factors in ways that we will describe next. We use our own assumptions about appropriate prices for inputs and outputs.

---

[1] SSEB, 2006. Our analysis uses case 3 from Appendix A, "Coal-to-Liquids Case Studies." Bartis, Camm, and Ortiz, 2008, described this plant's technological elements relevant to the cash-flow analysis offered here.

Figures of Merit

The cash-flow analysis considers real (adjusted for inflation) cash flows in 2007 dollars to and from the private investor and to and from the government over five years of investment to build a plant and 30 years of operating the plant.

It uses the real after-tax IRR associated with real cash flows to and from the private investor to measure the effects of these flows on the investor. We do not identify a hurdle IRR. Rather, we frame the analysis in the following way. (1) Given the nature of the investment, we believe that some investors would value early CTL production experience, which generates information that cannot be transferred in any formal way to parties not directly involved, enough to accept an IRR well below any hurdle normally applied to investments in an industrial plant.[2] In fact, we believe that the government could benefit from identifying such companies and favoring them in any process used to select investors for government assistance. (2) That point aside, we believe that any investor's interest in participating would increase as the real after-tax IRR that it anticipated rose. (3) We are most interested in circumstances that yield a midrange value of real after-tax IRR—say 5 to 15 percent—in which some policy instrument could move an investor from a clear decision not to invest to a decision to invest. We are especially interested in the relative cost to the government of using alternative financial-policy instruments to increase the real private after-tax IRR by one point in this range.

The analysis uses OMB's prescribed 7-percent real discount rate to calculate how real flows to and from the government affect the real NPV that the government associates with the project. The cost to the government rises as this NPV falls. OMB recommends accepting policies only if they are likely to yield a positive present value of net benefits, but it leaves some discretion to decisionmakers in individual agencies. That said, OMB (1992, §9) is not clear about how to treat high levels of uncertainty about NPV. It directs analysts facing uncertain outcomes to consider "key sources of uncertainty; expected value estimates of outcomes; the sensitivity of results to important sources of uncertainty; and where possible, the probability distributions of benefits, costs, and net benefits." It does not explain how to use such information to decide whether a policy is cost-effective. Nor does it describe how to use it when scoring the costs that an agency must associate with a policy with uncertain outcomes in its budget for the year the policy is approved for funding.[3] In practice, knowledgeable officials tell us that OMB tends to be conservative, initially demanding that new policies be scored to reflect the highest costs that might occur, but sometimes yielding unpredictably to agency pressure during negotiations.

In the face of considerable uncertainty, we focus on reporting the range of government outcomes we associate with any financial incentive package without asking specifically how the government would aggregate this information to decide whether the incentive package is cost-

---

[2]   Early commercial experience with CTL will work best to accelerate a viable global CTL industry if information on this experience is shared. Any federal program to support an early CTL project would presumably include requirements that the project generate information that would be made public. But the hands-on execution of such a project would also generate information that would be difficult to share, because it was latent in the task itself—perhaps captured mainly in the heads of the people who do it. Such latent information can be substantial, especially when a task is being refined. A basic principle of quality management is that those with hands-on experience with the task are usually the ones best qualified to improve it. Full understanding of a process typically requires deep interaction with those who perform it. Investors seeking to make a long-term commitment to the CTL industry could easily value the creation of such latent knowledge enough to accept a hurdle rate on a CTL project well below the rates they apply to other types of projects.

[3]   *Scoring* is a complex OMB process that determines how agencies can allocate their annual budgets in any year.

effective. As noted, we have a special interest in using information about real government NPV to compare the cost to the government of increasing the real private after-tax IRR in different ways. In our search for robust financial incentive packages—packages that reduce uncertainty about outcomes for a private investor—we give more attention to robustness in private IRR than in government NPV and tend to favor the central tendency of government NPV over extreme values when examining any one incentive package.

Relevant Decision Points

Over the life of the project, three important decision points frame the cash-flow analysis. The *first* decision point occurs before a front-end engineering and design (FEED) study generates detailed engineering and cost information on a site-specific plant. The decision in question is whether to initiate a FEED study, or continue one under way, in order to refine our understanding of the engineering and cost characteristics of a new CTL plant. That is, in effect, where we stand today in the United States; a few FEED studies are under way but not yet completed (Bartis, Camm, and Ortiz, 2008). The FEED study will reduce uncertainty about cost factors critical to the execution of the cash-flow analysis and the effects of policies that we study with it.

The *second* decision point occurs when the FEED study and associated financial analysis are completed. For the purposes of our cash-flow analysis, we assume that the FEED study resolves a significant range of uncertainty that currently exists about the costs of building and operating the plant. Only after this uncertainty is resolved does the government choose a final financial incentive package designed to attract private investors and choose a small number of investors to go forward. We expect the appropriate incentive package to depend significantly on the information revealed in the engineering study. When the government has chosen investors, investment begins. Our principal unit of observation is the set of cash flows associated with only one specific investor operating under one package of government policies at a time.

The *third* major decision point occurs five years later, when investment is complete and before operation begins. The costs of investment are now sunk. Looking forward, the investor decides whether to begin operation or terminate the project. Once operation begins, we assume for the purposes of this cash-flow analysis that operation continues for 30 years and then abruptly stops. Because the government wants both to build and to operate a plant to promote early CTL production experience, this third decision point places a constraint on the financial incentive package the government can offer. The package must be designed to ensure that, over the range of potential prices and costs that might prevail in the future, the private investor will choose not only to build the plant, but also to operate it.

Policy-Relevant Uncertainties

We examine the effects of each financial incentive package over all combinations of the following values of parameters:[4]

---

[4]   Our basic approach follows the lead of Lempert, Popper, and Bankes, 2003, using a transparently structured, simple model to explore the effects of various sources of uncertainty on policy outcomes. Our analysis led us to explore a narrower range of uncertainties than we initially anticipated, for reasons explained later. To achieve our specific analytic goals, we ultimately developed our own approach to explore the effects of uncertainty. Discussions with Paul Davis were invaluable during the design of this approach.

- Average price in real dollars per barrel of imported oil: $30, $35, $40, $45, $50, $55, $60, $65, $70, $80, $90. We assume that this price persists (in real dollars) over 30 years of operation. The low end of this range is consistent with recent low-oil-price projections (EIA, 2007). At world oil prices above $90 per barrel, the policy implications of government subsidies to early production experience are straightforward: Both government and private investors would reap large benefits, as we shall demonstrate next.
- CO_2-disposal cost in real dollars per ton: $0, $10. Our financial analysis is based on a CTL plant that captures 85 percent of the CO_2 that would otherwise be released into the atmosphere. Our capital and operating costs cover compressing and dehydrating this captured CO_2 so that it is ready for pipeline transport. The CO_2-disposal costs of $0 and $10 are used in a sensitivity analysis. For our base case, we assume that, over the plant's 30-year operating life, a third party is willing to transport from the plant all of the captured CO_2 at no cost to the CTL plant owners. In the initial operating years of early CTL plants, captured CO_2 can likely be sold to oil-field operators for enhanced oil recovery (EOR). However, national policy measures to reduce greenhouse-gas emissions are likely to make a large amount of CO_2 available, which would cause its price to fall to zero. For our sensitivity case, we assume that the project must arrange for the captured CO_2 to be transported to a sequestration site and permanently sequestered. We estimate that this will impose a project cost of no more than $10 per ton of captured CO_2 (Bartis, Camm, and Ortiz, 2008).
- Project costs: a reference case and high-cost case, as documented by Bartis, Camm, and Ortiz (2008). We assume that a FEED study will reveal one of these two cases as the correct case and then condition our analysis of incentive packages under the assumption that we will know which case is correct before the incentive package is finalized.

We do not assign subjective probabilities to these alternative futures or aggregate figures of merit across alternative futures to calculate any expected values of private IRR or government NPV. Rather, we sustain an awareness of the range of outcomes that might be associated with each policy we examine and seek policies that look good over these ranges of uncertainty.

We initially anticipated exploring variations in the prices of coal and electricity as well. But Energy Information Administration (EIA) projections anticipated little variation in these, relative to the variation anticipated for oil prices, over the next 30 years. And over the past 30 years, coal and electricity prices have not varied nearly as much as have oil prices. Limited explorations suggested that considering any reasonable range of uncertainties in these prices was unlikely to change our qualitative findings.[5]

Effects of Alternative Financial-Incentive Packages

With the effects of various combinations of policy instruments subject to the ranges of uncertainty just described, we explored the following parameter values:

---

[5] Hillard Huntington noted in a personal communication that, in all likelihood, substitution relationships would ensure that some positive correlation would exist between oil prices on the one hand and coal and electricity prices on the other over the life of any project. Any correlation would tend to reduce the variability that we identify in net cash flows from a project, because prices of inputs and outputs associated with the project would tend to move in the same direction when they change. If we built these relationships into our analysis, it would not yield as wide a variability in net cash flows as the analysis presented here. Of course, unknown unknowns—surprises that we do not know enough about even to speculate—could yield a wider range of variability in cash flows than that revealed by this analysis.

- fixed real price floor for all fuel produced, at oil-equivalent real price per barrel: $35, $40, $45, $50, $55, $60
- government guarantee to purchase all production at a fixed real oil-equivalent price per barrel: $60, $65, $70
- net-income sharing between the investor and government when the average price of imported oil rises above $60 per barrel: (1) none. (2) increasing government share as price rises. The government share implied by state and federal corporation income taxes is 0.36112. The shares used in our analysis rise linearly from 0.36112 at an average oil price of $60 per barrel to 0.41 at $70 per barrel to 0.46 at $80 per barrel, to 0.51 at $90 per barrel.
- investment credit for plant costs: 0 percent, 10 percent, 25 percent. In all cases, we assume that either (1) the investor has sufficient taxable income beyond this project to use all tax benefits generated by these instruments at full value as soon as they become available or (2) the credits are fully transferable.
- tax depreciation for nonland investment costs: (1) double-declining balance (DDB) with seven-year asset life,[6] and (2) 100-percent expensing in the first year of operation
- production credit in real dollars per barrel of oil-equivalent production: $0, $4, $8
- federal loan guarantee: (1) none. (2) guarantee for all debt; the guarantee effectively reduces the private investor's cost of debt by two real percentage points.

We used a spreadsheet that could easily accommodate many other financial-incentive packages to conduct the cash-flow analysis. We selected those listed here as representative of the options currently under discussion so that we could compare their effects under the range of uncertainties just described.

Desiderata

We seek packages that, at the second decision point described, (1) limit low private after-tax IRR rates and thereby reduce private risk by shifting it to the government, (2) limit private after-tax IRR rates well above those likely to induce participation in exchange for the government accepting the risk of low private IRRs, and (3) subject to the first two goals, limit the central tendency of total cost to the government. And to be considered desirable, a package must induce the private investor to operate the plant when investment is complete.

---

[6] This is the schedule that the tax code prescribes for a nonresidential investment in real property that has no other specific schedule assigned for its asset type. See IRS, 2007. Our analysis uses the schedule for an asset with a seven-year life in IRS, 2007, Table A-2, p. 72.

CHAPTER FOUR

Policy Effects with 100-Percent Equity Financing

This chapter uses the model described in Chapter Three to assess the effects of various packages of public policies when the investor uses 100-percent equity financing. Figure 4.1 presents baseline findings on which we will build through the remainder of this chapter. We will use the format in this figure repeatedly as we consider variations on this baseline. The figure shows real private after-tax IRR on the horizontal axis and real government NPV on the vertical axis. Dashed axes at a private after-tax IRR of 10 percent and a government NPV of zero offer benchmarks the reader can use to keep the results in perspective. As noted in Chapter Three, *we do not view these as meaningful hurdle rates for private investors or the government.*

Figure 4.1
Private and Government Effects with No Active Policies in Place: The Null Case

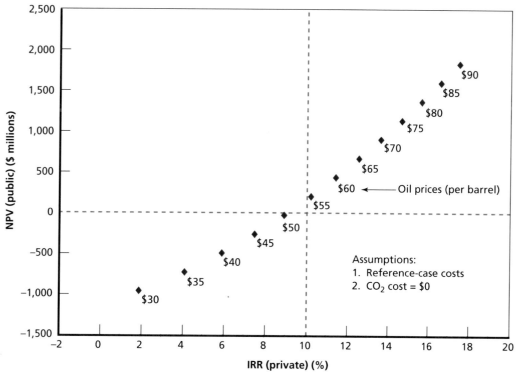

Policy-Relevant Sources of Uncertainty

Each point in this space presents the private after-tax IRR and government NPV associated with a particular financial-incentive package when oil and CO_2-disposal costs and investment and operating costs take certain values. The black diamonds in Figure 4.1 show oil prices (per barrel, as [IRR, NPV] pairs) for a situation in which (1) the government offers no special policies to encourage private-investor participation (2) the average cost of disposing of CO_2 is zero, (3) the reference case holds for project costs, and (4) average oil prices vary from $30 per barrel (yielding low IRR and NPV) to $90 per barrel (yielding high IRR and NPV). Potential variation in average oil prices alone yields a range of potential outcomes of about 2 to 18 percent in real private after-tax IRR and –$1.0 billion to $1.8 billion in real government NPV.[1] This kind of range is characteristic of the uncertainty that can reasonably be associated with any financial-incentive package. One, of course, can reasonably object that the range of average oil prices reflected here is far too wide for averages that hold over 30 years. Suppose we were to posit that a more reasonable range were $45 to $75 per barrel. Real private after-tax IRR still varies from 7 to 13 percent and real government NPV from –$270 million to $780 million for this null case. The figures of merit about which we care are highly sensitive to the average future price of oil.

Are they as sensitive to average CO_2 disposal costs? Figure 4.2 reports information for this question. Figure 4.2 shows the pairs from Figure 4.1 as the beginning points for the blue arrows. The blue squares at the heads of these arrows lie at pairs relevant to a CO_2 disposal cost of $10 per ton. Real private after-tax IRR falls by about 3.6 percentage points at an average fuel price of $30 per barrel. At higher average fuel prices, the change becomes progressively smaller until it reaches about 1 percentage point at an average fuel price of $90 per barrel. Real government NPV falls about $270 million at all average oil prices. Note that the IRR-NPV pairs move along exactly the same locus when CO_2-disposal costs fall or fuel prices rise; the two are directly linked in our model. A $10 per ton increase in the cost of CO_2 has a slightly larger effect on these pairs at all average fuel prices than does a $5 per barrel increase in fuel price. These effects are significant but small relative to the effects induced by any significant variation in average oil prices.

Figure 4.3 displays what happens when we add the effect of moving from the reference case to the high-cost case for project costs to the effect of increasing CO_2 costs by $10 per ton. It shows the baseline pairs from Figure 4.1 as black diamonds. The red arrows start at the blue boxes from Figure 4.2 and end at red triangles that show (IRR, NPV) pairs to a CO_2-disposal cost of $10 per ton and high-cost case project cost. Such a move cuts real private after-tax IRR by about 2.5 to 3 points and real government NPV by about $480 million at all oil prices. These are significant effects, but again small relative to the effects of changing average oil prices over even small ranges.

---

[1] NPV to the government can be negative in the absence of any subsidy, because the government gives up tax revenue when private cash flow from a project is negative and the project owner can use this loss to offset tax obligations that flow from net income elsewhere. At low oil prices, the project shown here does not generate high enough revenues over its lifetime to offset capital changes that the investor can use to reflect its initial investment in its tax calculations. This relationship between private net income and government tax revenues helps explain the positive relationship between private IRR and government NPV throughout this discussion. That is, if private IRR expected at the beginning of a project is so low that an investor will not commit to the project in the first place, the government will not suffer from negative NPV, because the project will never occur.

Figure 4.2
Sensitivities to Carbon Dioxide Cost in the Null-Policy Case

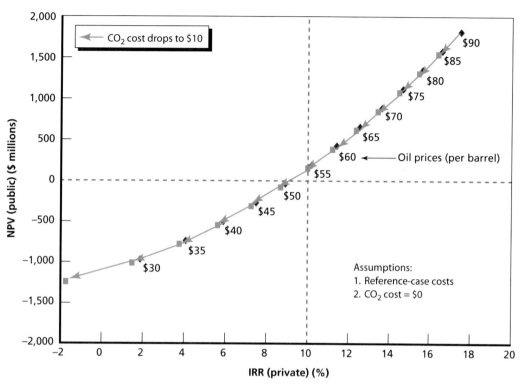

NOTE: For each oil price shown, the tail of the arrow shows the (IRR, NPV) pair for a zero cost of carbon dioxide, and the head of the arrow shows the (IRR, NPV) pair for a $10 cost of carbon dioxide. Each arrow shows the effect of changing the cost of carbon dioxide at a different assumed oil price.

RAND *TR586-4.2*

The slopes of the arrows tell us that, when real after-tax IRR and average fuel costs are both high, an increase in cost to the government hurts the investor more if it results from higher project costs than if it results from higher CO_2 costs. When real after-tax IRR and average fuel costs are both low, the converse is true: The same increase hurts the investor less if it results from higher project costs than if it results from higher CO_2 costs. That is because of differences in the timing of costs during the project life cycle and in the effective discount rates of the investor and the government. The investor becomes less and less sensitive to increases in the cost of CO_2, relative to the government, as the investor's effective discount rises with higher IRR. This difference gives us our first taste of a systematic difference between an investor and the government that, as we shall see, we can exploit to their mutual advantage.

In the remainder of our cash-flow analysis in this chapter, we focus on the effects of variation in average oil prices. Chapter Six revisits variations in CO_2 and project costs and shows that they are important to public-policy design.

Figure 4.3
Sensitivities to Carbon Dioxide Cost and Project Cost in the Null-Policy Case

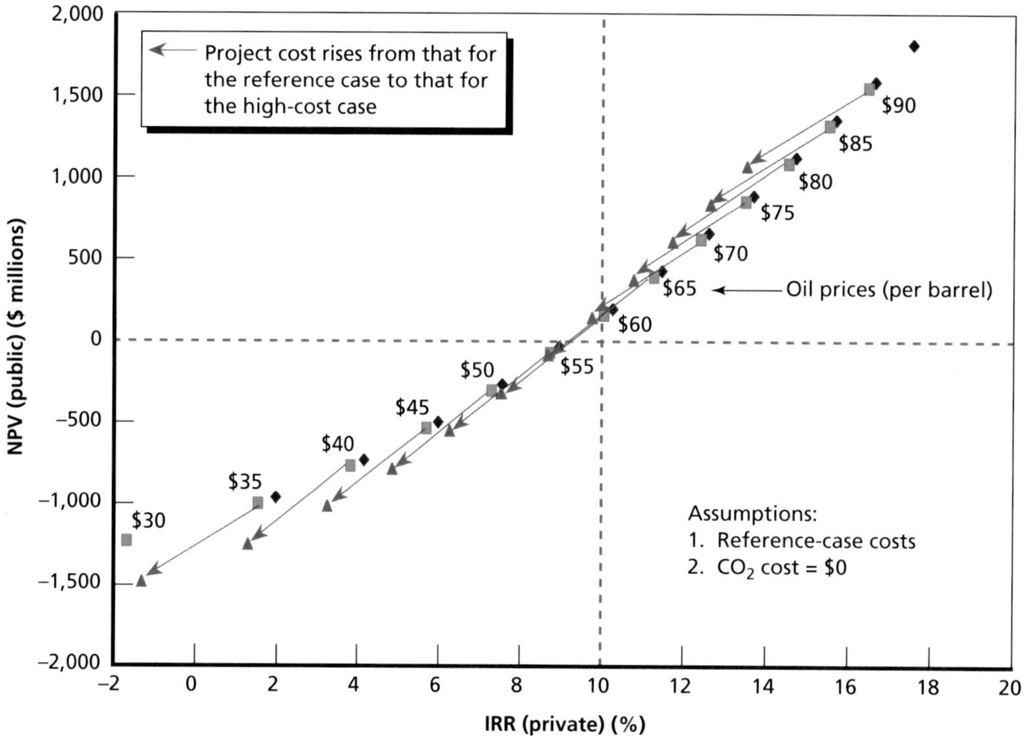

NOTE: At the tail of each red arrow (at the blue square), the carbon dioxide cost is $10, and the project cost comes from the reference case. At the head of each red arrow (at the red triangle), the carbon dioxide cost is $10, and project cost comes from the high-cost case. There is one arrow for each assumed oil price.

RAND TR586-4.3

Price Floors

Figure 4.4 explores the effects of introducing price floors as a financial incentive to encourage private-sector participation. In this particular case, we assume a price floor on all CTL-based fuel that the CTL plant produces. We need to be careful how we interpret that meaning of a price floor in the context of this model. In reality, as oil prices routinely fluctuate over time, a floor on oil prices would change the price that an investor received during any period in which fluctuations brought the price of oil below the floor. Even if average oil price never fell as low as the floor, the floor would affect periodic fluctuations below the floor, raising the average price that the investor received. Because this model holds the oil price constant over time, it cannot detect this kind of effect.

Imagine raising a legally stipulated price floor steadily through a range of oil prices fluctuating around a steady-state mean. As long as no fluctuation falls low enough to hit the floor, the floor is nonbinding and has no effect on the average price to the investor. But as the stipulated floor continues to rise, fluctuations fall low enough to hit it increasingly often, steadily raising the average price that the investor receives to above the legally stipulated price floor. As

Figure 4.4
Effects on Private Investor and Government of Progressive Increases in a Price Floor

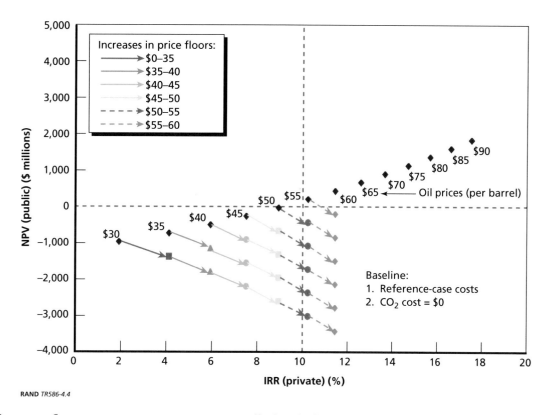

RAND TR586-4.4

the price floor continues to rise, it eventually binds for every fluctuation in oil price.[2] At that point, the average price that the investor receives equals the stipulated price floor. The result: The average price that the investor receives lies above the price floor until the floor rises high enough to affect every fluctuation in oil price.

In effect, the price floor used in this model corresponds to the *average price that the investor receives* when the floor is binding. As a result, the value of the price floor stipulated in policy that yields this average price received by the investor lies somewhat below the value used in the model. For example, the legally stipulated price floor that this model associates with a value of $40 per barrel lies below $40 per barrel—perhaps at $30–$35 a barrel. The exact distance depends on how volatile oil prices are at any point in time.

Figure 4.4 begins with the pairs shown in our null case in Figure 4.1. Arrows of differing colors now show the effects of successive increases in a price floor from $0 per barrel to $35 per barrel (red arrow), $40 per barrel (blue arrows), $45 per barrel (green arrows), and so on. The inset identifies additional increases. Several effects are worth noting:

1. The lowest levels of real private after-tax IRR disappear, progressively reducing the range of uncertainty about real private IRR, from the bottom up.
2. The lowest levels of real government NPV get lower, progressively increasing the range of uncertainty about real government NPV to include lower values. This trend ulti-

---

[2] No matter how high the price fluctuates upward, the price floor is so high that the floor remains binding for any fluctuation that occurs. It binds for a fluctuation. The floor binds at the price that occurs with every fluctuation.

mately begins to move the central tendency of real government NPV as well, but only well after minimum real private after-tax IRR has reached a level high enough to ensure private-sector participation. By itself, a price floor would not likely have to rise nearly this high.

3. The marginal cost in real government NPV of raising the real private IRR one point rises with each increase in the price floor. This is shown by the slightly increasing slope of the arrows associated with each increase. Table 4.1 displays the numbers underlying this pattern. The first column shows the increases in price floor represented by each set of same-color arrows shown in Figure 4.4. The second column shows the cost, in millions of dollars of real government NPV, of raising real private IRR by one percentage point. This cost rises from $188 million to $340 million as the price floor rises from $30 to $60 per barrel.

Later in this report, we will see more examples of how policy instruments can reduce the range of uncertainty about real private IRR, but almost always by increasing in the range for real government NPV. The likely relative willingness to bear risk suggests that such a change in ranges of uncertainty could be desirable unless it reduces the central tendency, however measured, of real government NPV too much. We will also examine how alternative financial-incentive packages affect the government's cost of raising real private IRR by one percentage point. For convenience, we will refer to this metric as the *government cost of raising IRR one point*, or CRIOP. We will call a CRIOP that varies when a single policy instrument is introduced or the incentive level is raised as a *marginal CRIOP*, which measures the additional cost to the government of each new increment of 1 point in real private after-tax IRR.

Income-Sharing Agreements

Figure 4.5 adds a policy instrument to the CTL-based-fuel price floors explored in Figure 4.4. The second policy instrument is an agreement to share net income between the private company and the government when average oil prices rise above $60 per barrel. The specific sharing agreement used in this analysis, described in Chapter Three, is purely illustrative, but it is compatible with real-world agreements that increase the government share of net income as

Table 4.1
Government Cost of Using a Price Floor to Increase Real Private IRR

Change in Price Floor ($/barrel)	Marginal Cost of Using Price Floor to Raise Private IRR ($ million per point)
30–35	188
35–40	228
40–45	261
45–50	291
50–55	317
55–60	340

the private investor's real IRR rises in response to higher oil prices (Kretzschmar and Kirchner, 2007).

The new arrows in Figure 4.5 show how our illustrative version of this instrument can dramatically limit real private IRR at high oil prices once a price floor has limited downside exposure at low oil prices. The dotted orange arrows show the effect of moving from no income sharing to the formal income-sharing agreement that we posit. By design, nothing happens under this agreement until the average oil price rises above $60 per barrel. At $60 per barrel and above, implementation of the income-sharing agreement transfers income from the private firm to the government; the amount transferred rises as the average price of oil rises but (by design) never enough to reduce the real private IRR at higher oil prices.

Table 4.2 uses a marginal version of the CRIOP—again, the incremental cost to the government of a one-point rise in real private after-tax IRR—to assess the cost-effectiveness of the transfer, at different average oil prices, achieved with this illustrative income-sharing agreement. The first column shows the prevailing average oil price in which the agreement operates. The second column shows the transfer to the government for income sharing. The third column shows the transfer from the private investor for income sharing. The last column shows the transfer to the government for each reduction in real private IRR by one percentage point, which is essentially a marginal CRIOP. A government decision to eliminate the illustrative income-sharing agreement shown here, to increase real private after-tax IRR by one point at any average oil price, would cost the government the amount shown in the table. These results

Figure 4.5
Effects on Private Investor and Government of a Price Floor with a Net Income–Sharing Agreement

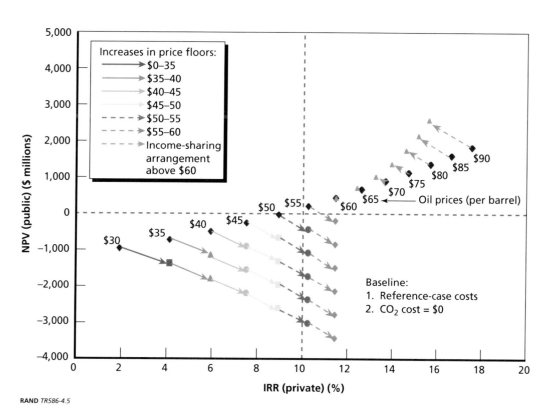

RAND TR586-4.5

Table 4.2
How Income Sharing Transfers Income from Investor to Government at Different Average Oil Prices

Average Oil Price ($/barrel)	Change in Government NPV ($)	Change in Private IRR (points)	Marginal CRIOP ($ millions/point)
65	46	−0.2	230
70	124	−0.5	248
80	375	−1.1	341
90	756	−1.9	398

also illustrate how the size of the pot to be shared grows with each increase in average oil price. Private after-tax IRR continues to grow with average oil prices despite the fact that the government gets more from each increase in real private IRR as the average oil price increases.

With no possibility of facing low prices, the private firm no longer needs the possibility of the large profits that would come with high prices to make participation attractive. In effect, when oil prices are high enough to ensure private participation even when sharing with the government, such a sharing agreement can compensate the government for the insurance it gives the private firm by removing the threat of low prices. This kind of arrangement becomes more attractive the larger the difference in risk aversion between the private firm and the government and the more likely concern is to arise about a government policy that leads to very high profits for a private firm.

As in Figure 4.4, and again by design, the incentive package in Figure 4.5 narrows the range of outcomes for the private firm while increasing the range of outcomes for the government, but having a much smaller effect on the central tendency of outcomes for the government.[3]

If we want to limit the range of outcomes for the private firm, a price floor can be raised and an income-sharing agreement tightened until the government ensures a fixed price for all CTL-based fuel production over the life of the project and the private firm agrees to accept that price, regardless of what average oil prices are outside the project. Figure 4.6 illustrates this approach for several levels of firm fixed oil prices. The pairs shown along the dotted line are the same as the pairs in our baseline in Figure 4.1.

A price floor of $55 per barrel would induce the $55 guarantee changes (red arrows). Qualitatively, these changes are equivalent to those shown in Figure 4.5. We can think of the changes shown here as simply carrying the changes in Figure 4.5 to the logical extreme, at which variation in real private after-tax IRR is driven to zero. The flexibility of using a price floor and income-sharing agreement together is likely to achieve a distribution of risk that dominates that induced by any firm fixed price. This flexibility is also better able to limit differences between prices inside and outside the project that could threaten the project's viability over the long term. Recall from our discussion of principles that such differences create more pressure for one party or the other to seek to change or exit the initial agreement as the differences get larger.

Once the instrument of a firm fixed price is chosen, a strict zero-sum relationship exists between the government and private firm. The blue arrows illustrate this by showing the effects of raising the price floor from $55 to $60 per barrel. Real government NPV falls by the same amount at any oil price; real private IRR rises by the same amount. As the green arrows, which

---

[3] *Central tendency* indicates the location of the middle of a statistical distribution, e.g., mean, median, mode.

Figure 4.6
Effects on Private Investor and Government of a Firm Fixed Price for Oil

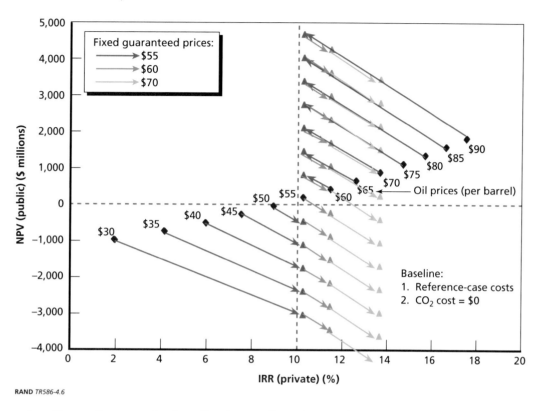

RAND *TR586-4.6*

show the effects of raising the firm fixed price from $60 to $70 per barrel, show, this zero-sum effect occurs at any level of a firm fixed price. In effect, it is present even in the initial decision to move from the baseline to any level of firm fixed price; the choice of a specific firm fixed price embodies within it such a zero-sum effect. This places the public and private interests in direct competition with one another, leaving no room for mutual gains once the firm fixed price instrument is chosen.

Investment Incentives

Once we accept a price floor and income-sharing agreement, we can then ask how additional instruments would affect outcomes for the government and private firm. Consider, for example, adding accelerated tax depreciation that replaces the DDB approach used in the baseline with immediate expensing during the first year of production. The green arrows in Figure 4.7 show the effects of such a change relative to a baseline from Figure 4.5 with a price floor of $40 per barrel. The effect moves resources from the government to the private firm at all levels of average oil price. The transfer decreases with decreasing average oil prices, creating the smallest incentives for private participation when they are needed most—at low average oil prices—and the largest incentives when they are needed least—at high average oil prices. These unwanted effects would be even worse without a price floor and income-sharing agreement. The price floor and income-sharing agreement can be seen in this context as allowing the use of immediate expensing to boost real private IRR in a middling range of average oil prices while moderating the unwanted effects of immediate expensing at low and high average oil prices.

Rather than simply accelerating tax depreciation, the government could give the private firm a direct grant with which to offset a portion of its investment costs. Figure 4.7 shows the effects of one approach to doing this, a 10-percent investment tax credit, with DDB depreciation over a seven-year project life for the company's share on investment cost. The red arrows show the effects of doing this at different average oil prices. Two points are worth noting.

First, at all levels of average oil price, the marginal CRIOP is higher for 100-percent expensing than for a 10-percent tax credit. We can see this by comparing the slopes of the arrows. The slopes are systematically steeper for expensing than for the tax credit. Table 4.3 displays the actual numbers underlying the figure. At various prices of average oil, marginal CRIOP is 2.5 to 4.0 times higher for 100-percent expensing than for an investment tax credit. This difference in cost reflects the difference in government and private-sector costs of capital; the tax credit is more cost-effective, because it transfers resources from the government to the private firm earlier.

Second, both options are more cost-effective for raising real private IRR when oil prices are middling or high than when prices are low. This pattern would be even stronger in the absence of a price floor. As a result, neither is well suited to addressing the goal of specially targeting low levels of real private IRR for enhancement. That said, once a $40 per-barrel price floor is in place, an investment tax credit is a more cost-effective way to raise real private IRR

Figure 4.7
Effects on Private Investor and Government of a Price Floor, Net Income–Sharing Agreement, and Alternative Investment Incentives

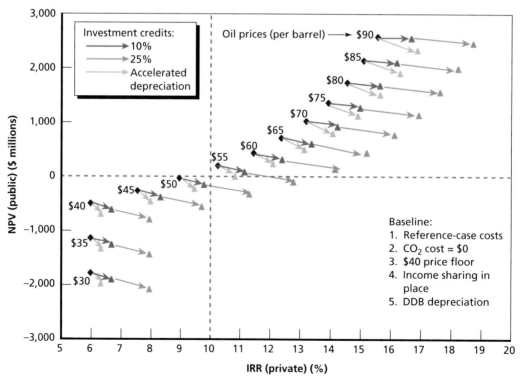

Table 4.3
Relative Cost-Effectiveness of 100-Percent Expensing and an Investment Tax Credit at Different Average Oil Prices

Average Oil Price ($/barrel)	Marginal CRIOP for 100% Expensing ($ millions/point)	Marginal CRIOP for 10% investment tax credit ($ millions/point)	Relative Marginal Cost to Raise Private IRR[a]
30	673	170	4.0
40	673	170	4.0
50	337	133	2.5
60	337	133	2.5
70	256	102	2.5
80	235	76	3.1
90	220	59	3.7

[a] This is an absolute number. It is the ratio of one number (measured in dollars per point) to another number (measured the same way). So, if one item costs $10 and another costs $5, the ratio of the two is 2—the first costs twice as much as the second, per item.

one point than increasing the price floor to do the same thing at prices somewhere above $40 per barrel. A more aggressive income-sharing arrangement could be used to limit excessive incentives—incentives beyond those required for private participation—induced by a higher investment tax credit at high oil prices.

Taken together, these two observations suggest a strong preference for using a tax credit over using expensing to achieve any increase in the incentive for private participation. A third observation adds another reason to favor the tax credit. In Figure 4.7, the blue arrows show the effect, at each level of average oil price, of increasing the investment tax credit from 10 percent to 25 percent. Doing this allows a marked increase in the incentive for private participation with little change in the marginal cost to the government of doing so; that is, the slope of the blue arrows is only slightly lower than that for the red arrows, despite a large expansion in the size of the tax credit. This gives the government more flexibility in tailoring an investment tax credit to the mutual preferences of government and private firm than is possible with immediate expensing of private investment costs.

The blue arrows also illustrate that the incentive created for private participation remains higher at higher average oil prices than at lower oil prices, an undesirable feature by itself. Once we have chosen a tax credit over expensing, however, an incentive package can be crafted that combines a more aggressive tax credit to increase the incentive for private participation at all oil prices with a more aggressive income-sharing agreement, which limits the undesired effects of the increased investment tax credit at higher oil prices, when government encouragement is not necessary. Again, a combination of policy elements allows us to craft a package of elements that advances the mutual interests of the government and private firm.

Production Incentives

Rather than adding investment incentives, we could add production incentives to a price floor and income-sharing arrangement. The red arrows in Figure 4.8 show the effects of adding a

Figure 4.8
Effects on Private Investor and Government of a Price Floor, Net Income–Sharing Agreement, and Production Incentives

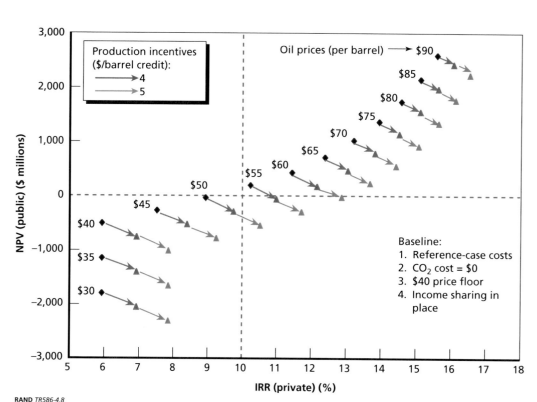

RAND TR586-4.8

simple $4 per-barrel tax credit (subsidy) to the same baseline package used in Figure 4.7. Effects are larger the lower the average oil price, an inherently desirable characteristic. The blue arrows show the effect of expanding the production tax credit from $4 to $8 per barrel. Because the slopes of the red and blue arrows at any average world oil price are very close, the CRIOP for a production credit at any single average world oil price is fairly constant. Table 4.4 confirms this. The two numbers in the third column are the CRIOP values calculated for moving the production tax credit, respectively, (1) from $0 to $4 per barrel and (2) from $4 to $8 per barrel. As average oil price rises, however, the slopes of the red and blue arrows increase, showing that a production tax credit grows less cost-effective as average oil price rises.

How cost-effective does a production tax credit look relative to an investment tax credit? Table 4.4 reports information relevant to this question. Two patterns are worth noting. First, at any average oil price, an investment credit is more cost-effective than a production credit. Again, this effect occurs because the investment credit transfers resources from government to investor early in the project, thereby taking advantage of the difference in government and private-sector costs of capital. Second, an investment credit grows rapidly more cost-effective as average oil price rises.

Given these results, a production credit can be attractive only if it is needed to ensure that the private firm decides to operate the plant after it has built it. That is not a problem with the project we use in this analysis. Under all assumptions in the analysis, the net annual cash flow to the private firm is positive after the plant is built. So there is no reason to consider the use of a production tax credit for such a project.

Table 4.4
Cost-Effectiveness of an Investment Tax Credit and a Production Tax Credit at Different Average Oil Prices

Average Oil Price ($/barrel)	Marginal CRIOP for 10% Investment Tax Credit ($ millions/point)	Marginal CRIOP for Production Tax Credit ($ millions/point)	Relative Marginal Cost to Raise Private IRR[a]
30	170	253–276	1.5–1.6
40	170	253–276	1.5–1.6
50	133	314–326	2.4–2.5
60	133	358–374	2.7–2.8
70	102	379–397	3.7–3.9
80	76	396–413	5.2–5.4
90	59	413–415	7.0

[a] This is an absolute number. It is the ratio of one number (measured in dollars per point) to another number (measured the same way). So, if one item costs $10 and another costs $5, the ratio of the two is 2—the first costs twice as much as the second, per item.

Summary

When an investor plans to use 100-percent equity financing, the most cost-effective way for the government to raise real private IRR one percentage point is to use an investment incentive, such as an investment tax credit (see Table 4.5). This incentive dominates 100-percent expensing of investment costs and production incentives at every average oil price. Unfortunately, any given increase in an investment tax credit increases real private IRR the most at higher average oil prices, when an increase is least needed.

This can be resolved in either of two ways. First, a modest price floor can boost real private IRR when it is needed most without raising private IRR otherwise. As the price floor rises, however, it grows increasingly costly relative to an investment tax credit. Second, a net income–sharing agreement can be used to offset excessive effects of an investment tax credit when average oil prices are high. This will be even easier to justify in the presence of a price floor that protects investors from the ill effects of low average oil prices. An investment credit has such a low cost at high oil prices (see Table 4.5) precisely because a net income–sharing agreement returns revenues to the government. As the amount returned rises at higher oil prices, the effective cost of the investment incentive drops rapidly. It would drop more slowly in the absence of such an agreement. The numbers in Table 4.5 show that a well-integrated financial incentive package, which would presumably include a price floor, income-sharing agreement, and investment incentive, can work well.

Great caution is required when interpreting the costs in Table 4.5. Because each column typically envisions a somewhat different financial incentive package, we cannot compare the cost numbers directly. For example, the costs of alternative price floors are measured in the absence of an investment incentive. The costs for the investment incentive envision an explicit $40 price floor. In the absence of such a floor, the costs of the investment incentive would be higher at oil prices below $40 per barrel. Similarly, the costs of the investment incentive envision a net income–sharing agreement. These costs would be higher at oil prices above $60 in the

Table 4.5

Summary of Marginal CRIOPs of Alternative Instruments with Equity Financing

Average Oil Price ($/barrel)	Instrument ($ millions/point)				
	Price Floor	Income Sharing[a]	100% Expensing	Investment Credit	Production Credit
30	188–260	NA	673	170	253–276
40	261–300	NA	673	170	314–326
50	318–329	NA	337	133	379–397
60	NA	NA	337	133	413–415
70	NA	248	256	102	253–276
80	NA	341	235	76	314–326
90	NA	398	220	59	358–374

[a] For practical purposes, this cost shows how much money flows from the government when the government abandons net income–sharing to increase the investor's real IRR by one percentage point. Note that, in Table 4.2, the government imposes income sharing; here, the government removes income sharing. These actions are functionally equivalent for the purposes of the numbers reported in Table 4.5.

absence of such an agreement. The costs for the agreement itself are measured in the absence of an investment incentive and so are not directly comparable to the costs shown for the investment incentive.

CHAPTER FIVE

Policy Effects with Debt Financing

The potential for using debt financing affects our analysis in two ways. First, for any particular set of policy instruments, it immediately leads to the potential for higher levels of real private IRR at any level of real government NPV—for policies with lower CRIOP levels. Second, it opens the door for government-provided loan guarantees for any loans an investor uses. We consider each effect in turn and then examine their combined effect on the outcomes associated with other policy instruments available to the government.

This chapter starts by showing how IRR rises when debt share increases. It then presents a loan guarantee as one of a variety of policy instruments that the government can use to loosen the resource constraints that lenders place on borrowers. One way is to reduce the cost of debt capital at a level of debt share. The second is to open the door for a borrower to choose a higher debt share. The chapter addresses each of these in turn. Unfortunately, the behavioral factors relevant to these effects and the influence of factors specific to any project are so complex that we cannot generate results based on cash-flow analysis for debt financing as we did in Chapter Four for 100-percent equity financing. Our various efforts to do so should serve as a warning about the difficulty of assessing, with confidence, the effects of policy under debt financing. Appendix B addresses these issues in more formal mathematical terms. This approach provides a more coherent way to view debt financing and the most important effects of loan guarantees than cash-flow models allow. This chapter also presents an empirical assessment of how a particular government loan guarantee program led to the bankruptcy of a major firm to illustrate the principles that emerge from the mathematical model.

How Debt Financing Affects Real Private After-Tax IRR

Debt financing can dramatically increase the real private after-tax IRR that an investor associates with a project. Figure 5.1 illustrates this point. The red arrows show the effects of moving from the equity-financed baseline we showed in Figure 4.1 in Chapter Four to a new situation in which an 8-percent (real) loan finances 40 percent of the private company's investment. We choose a relatively high real cost of debt capital (1) to reflect the degree of risk that should be associated with plants that provide early CTL production experience and (2) to make an a fortiori argument that debt financing can dramatically increase private after-tax IRR even if debt capital is relatively costly. Three patterns are worth noting in Figure 5.1:[1]

---

[1] When a policy change drives real private after-tax IRR to a negative value, our model cannot track that effect further. As a result, in Figure 5.1, we do not track the effects of debt financing on IRR when the average oil price is $30 per barrel.

Figure 5.1
Effects on Private Investor and Government of Increasing the Share of Debt in Project Financing

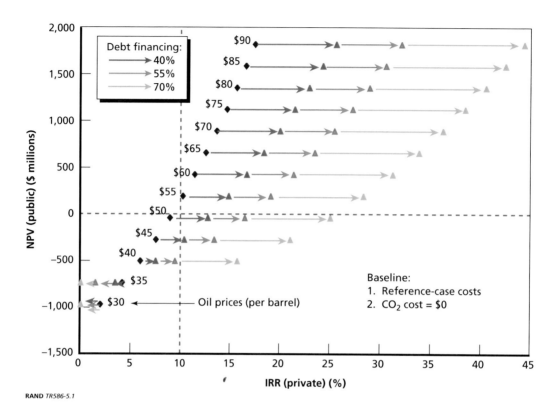

1. This change has no effect on the government, because the government's effective participation is not affected. Taxes previously paid on equity income are now paid by the lenders, whom we assume face the same tax rates as the investor in the plant does.

2. The change dramatically increases the real private after-tax IRR at most average oil prices. This effect increases as average oil price increases, creating the largest effects where they are least needed for public-policy purposes. That is because, as long as a loan's real after-tax interest rate is lower than the real after-tax private IRR with 100-percent equity financing, the loan allows the investor to reduce its investment costs more than its loan repayments rise when the investor evaluates its private NPV for the loan at its IRR for the project.[2] This change by itself can make an unconventional-liquid-fuel project profitable without any government support for all but the lowest average oil prices under review here.

3. A loan makes the investor worse off when its real after-tax IRR with 100-percent equity financing is lower than the real after-tax interest rate on the loan. We need to be cautious about how we interpret this finding. It reflects an assumption that the cash-flow analysis captures all costs and benefits relevant to the players. This is true for the lender; it is not for the investor or borrower, which may expect learning benefits from building

_____

[2] When an investor can deduct interest payments from its taxable income, the after-tax value of an interest rate differs from its pretax rate.

and operating the plant, beyond any financial benefits it gains from the project. That is, the investor can accept a project when the IRR we measure in this analysis is well below its actual equity cost of capital, precisely because additional investor benefits exist outside the scope of the cash-flow analysis. Only when the investor's debt cost of capital exceeds its true cost of equity investment does private after-tax IRR fall with debt share. That was not a relevant concern until debt entered the analysis. We now need to keep in mind this problem in our measure of private IRR as we interpret the results of the cash-flow analysis.

Viewed naively, the strong effect of debt financing on real private IRR potentially makes debt financing look irresistible. As the blue (moving from 40 to 55 percent debt financing) and green (55 to 70 percent debt financing) arrows in Figure 5.1 show, there is no end to this effect.[3] In principle, an investor would try to reduce its equity share as much as possible to increase its real IRR. But increasing debt share can increase the probability of default on the loan. This threatens income flows for the borrower and lender; the lender responds by increasing its interest rate to pass this increased risk back to the borrower.[4] The effects of higher debt share on real private IRR shown in Figure 5.1 do not reflect the effects of these predictable changes. To be complete, any analysis of a debt-financed project needs to make the probability of default and real cost of debt capital to the borrower contingent on the debt share the borrower chooses.

A word of caution is appropriate in interpreting the borrower's real cost of debt capital. As elsewhere in this analysis, it does not include inflation in the cost stated. In addition, it is real to the borrower because it reflects what the borrower expects to pay in return for a loan. If the borrower faces an 8-percent interest rate for a loan but expects only a 75-percent probability that it will actually repay the loan, the real cost of debt capital the borrower faces is $0.75 \times 0.08 = 6$ percent. Under these circumstances, the borrower would face a real cost of debt capital of 8 percent if it received a loan with a $0.08/0.75 = 10.56$-percent interest rate. In the discussion that follows, when we refer to *the real cost of debt capital*, we mean the expected cost associated with actually repaying the loan. This applies to the borrower and lender. For simplicity, we assume that borrower and lender share the same expectation of repayment and so associate the same real cost of debt equity with any quoted interest rate.[5]

How debt share affects the probability of default depends on the particulars of each case. Table 5.1 uses one case from our spreadsheet analysis to look more closely at the power of debt financing. It assumes that project costs reflect the high-project-cost case and that no government incentives are in place. With an average oil price of $55 per barrel and 100-percent equity financing, the real private after-tax IRR for the investment in these circumstances is 7.6 percent. Table 5.1 shows how four different measures of IRR change as debt share increases. They reflect the following two choices:

---

[3] We give particular attention to debt shares of 55 and 80 percent in this analysis and the analysis to follow in this report to allow comparability with analysis in a recent National Energy Technology Laboratory report, which emphasized these values in its analysis of the effects of a loan guarantee. See NETL, 2007, esp. p. 56.

[4] Appendix B explains in more detail the factors that shape this interaction.

[5] Appendix B provides more detail on how lenders' and borrowers' perceptions of risk affect their decisions.

Table 5.1
How Debt Share Affects Real After-Tax Private IRR

		Amount					
	Debt share	0	0.4	0.5	0.6	0.7	0.8
Adjustment	**Probability of default**	0	0	0.033	0.067	0.133	0.200
Loan payment fully adjusted to reflect probability of default	Annual loan payment ($/barrel)	0	4.15	5.36	6.66	8.38	10.38
	Annual equity payment ($/barrel)	12.34	9.68	8.91	8.08	6.98	5.70
	Private IRR, cash flows adjusted for probability of default (%)	7.6	10.6	11.7	12.8	13.9	15.8
	Private IRR, cash flows not adjusted for probability of default (%)	7.6	10.6	12.1	13.9	16.1	19.9
Loan payment not adjusted to reflect probability of default	Annual loan payment ($/barrel)	0	4.15	5.18	6.22	7.26	8.30
	Annual equity payment ($/barrel)	12.34	9.68	9.02	8.36	7.69	7.03
	Private IRR, cash flows adjusted for probability of default (%)	7.6	10.6	11.8	13.4	15.0	18.5
	Private IRR, cash flows not adjusted for probability of default (%)	7.6	10.6	12.3	14.4	17.8	24.6

- Whether the lender changes the interest rate to maintain a real expected return on the loan as the rising debt share increases the probability of default. The top half of the table assumes that the lender makes this adjustment; the bottom half assumes that the lender does not.
- Whether the cash flows used to calculate IRR reflect the rising possibility that all cash flows will fall to zero at some point as debt share rises. One row in each half of the table adjusts the cash flows; the following row does not.

Different columns of the table show each of these four versions of IRR for different debt shares.

As the debt share rises, the size of the loan taken increases, increasing the size of each annual repayment. In addition, beyond some point, the probability that the project will not generate enough net income to repay the loan increases as well.[6] The cost per barrel produced of repaying a 6-percent loan is shown in rows labeled "Annual loan payment." The annual cost of repayment rises with both size of loan and probability of default. Deducting the loan payment from the net income available to cover capital costs yields the "Annual equity payment" shown in the rows that follow the rows for loan payment. Despite the fact that this annual payment to equity falls as debt share rises, the IRR for private equity rises with debt share, because the amount the private investor must invest falls with debt share.

For our purposes, the most appropriate definition of IRR assumes that (1) the lender adjusts the interest rate to hold the expected loan payment constant as debt share rises and (2) the borrower adjusts its expected cash flows downward as a rising debt share increases the probability of default.[7] Using this definition, increasing the debt share from 0 to 80 percent increases the borrower's IRR from 7.6 to 15.8 percent. This is not nearly as dramatic an effect as that shown in Figure 5.1, in which the analysis takes no account of the points addressed in Table 5.1. But this doubling of IRR is still dramatic proof that debt financing can markedly improve a project's performance for a borrower.

Failures to reflect effects of potential default on the lender's loan rate or the borrower's expected cash flows lead to larger effects of debt share on private IRR. As Table 5.1 shows, these failures can allow real after-tax IRR to rise to anywhere from 18.5 to 24.6 percent when the debt share reaches 80 percent. Like the measure of IRR used in Figure 5.1, these measures are misleading.

By way of contrast, consider the effects on IRR of simply reducing the cost of debt capital without changing the debt share. Figure 5.2 shows the effect of reducing the cost of debt capital from 8 to 6 percent. It considers the circumstances in Figure 5.1 when 55-percent debt financing applies. The green arrows display the effect as the change moves the borrower

---

[6] To determine the probability of default, we assumed a uniform distribution of average oil price between $40 and $70 per barrel. At low average oil prices, it becomes increasingly unlikely that the investment will generate enough net income to cover loan repayment as the debt share increases. We offer this connection between debt share and probability of default entirely as an illustration.

[7] As Appendix B explains, the lender will probably raise the interest even more than this. But we have no empirical basis for determining how much more as the lender's perceived risk rises.

Figure 5.2
**Effects on Private Investor and Government of Reducing the Cost of Debt Capital Through a
Change in the Private Capital Market**

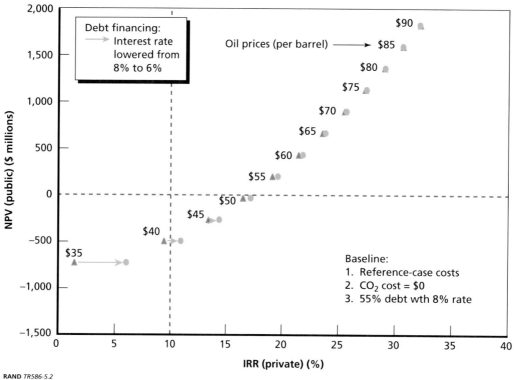

RAND *TR586-5.2*

toward those circumstances.[8] The effect is large when the average price of oil is low. Because of complex tax effects on our model, it falls to near zero at high average oil prices.

Note how improbable it is that the debt share would be the same at all average oil prices. If both the borrowing investor and the lender believe that they can predict the future oil price, the investor will face a lower default risk at higher average oil prices and hence will opt for a higher debt share when it expects higher average oil prices. Even if the investor is uncertain about future average oil prices, some ability to forecast future prices will lead to a positive relationship between average oil prices and debt share, a relationship that is not as strong as it would be with perfect foresight. Assuming no relationship effectively says that investors have no information about the future.

This becomes even more improbable when we imagine a future 30 years with multiple opportunities to revisit the financing for a project as information accumulates about average oil prices over this period. If expected future prices fall over the period, the investor may have difficulty adjusting its debt share downward to achieve its preferred level of risk. But if expected future prices rise over the period, the investor will respond by raising its debt share. These asymmetric opportunities to change debt shares over 30 years will encourage any investor to be more conservative in its choice of debt share early in production and to adjust the share upward if expected future prices rise. It will make a risk-averse investor relatively more conservative if it

---

[8]   Again, because the relevant IRRs are negative when the average oil price is $30 per barrel, we do not show results for this price.

expects low future average oil prices than if it expects high future average oil prices, inducing an ex ante positive correlation between the debt shares that investors choose initially and the future average oil prices they expect. Ex post, it will lead to a higher realized debt share over the 30-year period at higher realized average oil prices.[9]

In sum, unlike the case of 100-percent equity financing when debt share is zero at every average oil price, assuming the same level of debt share at every future average oil price is artificial when debt financing is available. But assigning appropriate levels of debt share to different average oil prices is complicated not only by the project-specific concerns already noted, but also by concerns about how an investor accumulates information about average oil prices in the future. Any effort to use cash-flow analysis to examine policy effects, such as those examined in Chapter Four, in the face of uncertainty about oil prices and other factors would require strong assumptions that we lack the theoretical or empirical basis to make with any confidence.

For now, we conclude that debt financing is attractive if the cost of debt capital is lower than the effective cost of equity capital. And once this condition exists, the primary power of debt financing comes from increasing debt share, not reducing the cost of debt capital per se.[10] More broadly, the preceding discussion should serve as a warning that the quantitative results of any cash-flow analysis of debt-financed investment in a CTL production capacity should be viewed with great skepticism and care.

Behavioral Effects of Debt Financing and Loan Guarantees

This section puts into words much of what Appendix B presents mathematically. Debt financing allows an investor in a project to increase its real after-tax IRR as long as the cost of debt capital available to the investor is below the IRR for the cash flows generated by the project. An investor can increase its real after-tax IRR by increasing the debt share as long as a higher debt share does not excessively increase the risk of default. Default risk ultimately rises with rising debt share because, given any realized profile of project cash flows, the cost of servicing increasing debt can ultimately absorb all of the project's realized cash flow and more. Recognizing this, lenders generally expect higher default rates for projects financed with higher debt shares. In response, all else being equal, lenders charge higher interest rates when investors maintain higher debt shares and ultimately limit how much debt capital they will offer for a project. Given how lenders behave, an investor chooses its debt share to balance the advantage of more low-cost debt capital with the disadvantage of higher default risk.

To the extent that a risk of default exists, a loan guarantee effectively allows a lender to offer loans to an investor at a lower rate by shifting the risk associated with default from the lender to the government. If no such risk exists, a loan guarantee will have no effect. The larger the default risk, the more a loan guarantee will reduce the rate a lender offers the investor. That is, a loan guarantee is more powerful precisely when it imposes a larger expected cost on the

---

[9] This discussion emphasizes the likely link between future average oil prices and debt share. Oil prices are not the only factors that influence the potential for default when an investor uses debt financing. We could just as easily expect a link between lower project costs or average CO_2-disposal costs on the one hand and a higher debt share on the other, ex ante and ex post, for exactly the same reasons.

[10] Appendix B addresses in more detail how a borrower might choose between seeking a lower interest rate and seeking a higher debt share when a lender allows a choice.

government agency offering the guarantee. In this regard, the guarantee acts very much like a direct government subsidy. But it is less visible than a direct subsidy when the guarantee is offered, because the probability that the government will incur a cost is less than one.[11]

A loan guarantee can encourage an investor to pursue early CTL production experience in two ways. First, if the investor maintains its debt share, a loan guarantee reduces the cost of debt capital to the investor. By reducing the investor's negative cash flows over the life of the CTL project, the loan guarantee increases its real after-tax IRR, as shown, for example, in Figure 5.2.

Second, a loan guarantee encourages an investor to increase its debt share. This can occur in two ways. First, since a lender no longer faces as much risk from a default when a loan guarantee exists, the lender is less likely to charge more for loans or to limit the size of a loan when the investor seeks to increase its debt share. Because the investor's cost of debt capital does not rise as much when it seeks a higher debt share, the investor chooses a higher debt share if it can. If the government agency guaranteeing the loan imposes the same close oversight that the lender would have provided before the loan guarantee, this source of encouragement should disappear. Second, even if the government provides close oversight, the cost of debt capital that the investor faces with its preguarantee debt share has fallen. As a result, at that debt share, its risk of default has fallen because, for any realized profile of project cash flow, it can now successfully service more debt. In response, the investor may decide to expand its debt share to balance more effectively the costs and risks that it faces.

This second source of encouragement to expand debt share exists precisely because the government's willingness to bear a portion of the risk of default effectively reduced the investor's real cost of debt capital. In other words, when it agreed to bear some risk of default, thereby allowing the interest rate on the guaranteed loan to fall, the government created an incentive for the investor to increase its debt share, increasing the probability of default, and so increasing the cost to the government of offering the loan guarantee.

The quantitative size of these effects—(1) how much an investor would increase its debt share in response to a loan guarantee or (2) how much that increase in debt share would increase the government's cost of the loan guarantee—varies from one project to the next. In the next section, we present information from an empirical analysis of one real loan guarantee that had available far more detailed information than we attempted to generate in our cash-flow analysis. We did not explore uncertainties associated with project performance that would allow us to estimate the size of these effects for the project examined here. But our analysis does yield three major findings:

- Except at very low expected petroleum prices, if the investor holds its debt share constant, a loan guarantee has only small effects on real after-tax IRR flows. Its effects on real after-tax IRR grow rapidly as the loan guarantee induces the investor to increase its debt share. In the end, a loan guarantee is likely to encourage investor participation in a project mainly by allowing it to benefit from the effects of using a higher debt share, not through its direct effects on the cost of debt capital.
- How much a loan guarantee costs the government depends fundamentally on how much responsibility the government takes to oversee the project in the same way that a private-sector lender without a loan guarantee would have. The more closely the government

---

[11] To see this demonstrated most directly, see Equation B.11 and Table B.2 in Appendix B.

manages the loan and limits the investor's debt share and hence its probability of default, the smaller the expected cost to the government (but also the smaller the positive effect of the loan guarantee on real private after-tax IRR).

- The power of any loan guarantee ultimately lies in how much default risk the government is willing to accept. As noted, a loan guarantee offers an investor no benefits unless a default risk exists. And, no matter how all these arguments resolve themselves in the financing of any particular project, a loan guarantee encourages more investor participation in the project the more the government is willing to subsidize the project by accepting more default risk.

The federal government should use a loan guarantee to promote early CTL production experience only with its eyes wide open about how that guarantee works, how much it is likely to cost the government to encourage investor participation to any degree, and how well the government can put in place an effective project-monitoring and control system capable of protecting the federal purse.

An Illustrative Example of Investor Decisionmaking Under Government Loan Guarantees

A useful example of how loan guarantees can affect investor decisionmaking is the failure of the Canadian Northern Railway (CNR). The use of Canadian federal and provincial loan guarantees and CNR's subsequent failure are exceptional only because they have been the subject of an extremely careful and detailed empirical analysis. By piecing together detailed historical data from company documents and using a model of optimal capital structure drawn from finance theory to place the data in perspective, Lewis and MacKinnon (1987) demonstrated that the loan guarantees offered to save the railroad from bankruptcy probably actually caused its failure, for precisely the kinds of reasons we have discussed.

In the years leading up to World War I, CNR was attempting to build a transcontinental railroad. Starting in 1912, it faced increasing difficulty raising capital and concluded in 1914 that it would go bankrupt without new capital that it could acquire through loans only if the government guaranteed them. In response, the federal government guaranteed $45 million in loans.[12] CNR expanded its debt. Additional provincial and federal loan guarantees followed, reaching $212 million by 1916. CNR expanded its debt further. Bankruptcy came in 1916, and the government ultimately paid all the railroad's debts, worth $418 million.

Lewis and MacKinnon (1987) used a model that estimated the optimal capital structure for CNR, from its owners' perspective, with any particular level of government loan guarantees. They show that, as the level of government loan guarantees rises, the firm's optimal level of debt rises, the firm's willingness to make incremental investments that could lead to bankruptcy rises, and, as a result, the ex ante probability of bankruptcy associated with those incremental investments rises. Using a range of assumptions, they estimated that CNR's ex ante probability of bankruptcy in 1916 was three to four times what it would have been in the absence of government loan guarantees. Because of the loan guarantees, it was in the CNR investors' interest to pursue incremental investments even if they associated an ex ante bank-

---

[12] All values are stated in then-year dollars. Stated in terms of today's dollars, these values would obviously be higher.

ruptcy probability as high as 69 to 74 percent with these investments. Put another way, the loan guarantees induced the investors to undertake incremental investments that they would never have undertaken in the absence of loan guarantees.

Lewis and MacKinnon's (1987) careful analysis indicated that, by insulating the firm and its creditors from the negative consequences of making bad decisions, Canadian government loan guarantees ultimately induced CNR to take on too much debt and to undertake unacceptably risky investments. These decisions drove CNR into bankruptcy. Why? Because none of the many government entities providing loan guarantees maintained the fiduciary responsibility that an investor would normally have expected from its owners and creditors, parties shielded from harm by the loan guarantees. This scenario has occurred over and over, because the government entities that provide loan guarantees rarely have the capabilities normally required to sustain effective fiduciary responsibility. Again, the CNR case is exceptional mainly because Lewis and MacKinnon made the effort to collect the data required to document investor behavior under a fairly unexceptional set of loan guarantees.

How Debt Financing Affects the Use of Other Policy Instruments

To the extent that any financial policy instrument affects the level of risk that a lender associates with a project, applying the instrument will tend to affect the cost of debt capital and the debt share for the project. For example, price floors clearly limit risk at low prices and so should reduce the cost of debt capital and increase the debt share expected at oil prices below a price floor; they should have no effect at oil prices above a price floor. An income-sharing agreement, on the other hand, could increase risk at high average oil prices by reducing the investor's cash flow when those prices are reached. Unless the arrangement was carefully designed to ensure the investor's success at high prices, it could increase the cost of debt capital and reduce the debt share expected at high oil prices. And the other instruments we have considered, by enhancing the project's real private after-tax IRR, would presumably reduce the project's cost of debt capital and increase the debt share associated with it at any average oil price. We cannot quantify these effects, but the preceding analysis makes it clear that these effects could be significant and deserve more attention when debt financing appears to be an important factor in an unconventional-fuel plant project.

CHAPTER SIX

Implications for Robust Financial-Incentive Packages

Using the spreadsheet underlying the cash-flow results described in Chapter Five, we can scan the outcomes associated with that range of average oil prices and ask how various financial-incentive packages affect private IRR and government NPV in two cases. In case A, project costs match the reference case, and CO_2 management imposes no costs other than those already assumed—namely, the costs for compression and dehydration of captured CO_2. In case B, project costs match the high-price case, and CO_2 costs $10 per ton to transport and sequester. Scanning these packages reveals a small number of packages that place private IRR in a moderate range for all prices of oil and likely prices for managing CO_2 emissions. In particular, we seek packages that are likely to place private real after-tax IRR between 6 and 15 percent when average oil prices lie between $40 and $70 per barrel over the next 35 years.[1]

Changing the criteria for choosing robust policy goals would obviously change the packages. Moving the private IRR range, say, to 10 to 20 percent would favor packages with higher price floors and investment incentives and less reliance on an income-sharing arrangement. Shifting the focus to a higher range of average oil prices would have the opposite effect. As noted, we expect different investors to have different preferences about robustness and hence the outcomes on which we should focus to design robust financial-incentive packages. The packages described here are illustrative and should not be interpreted as the right packages for the market relevant to early commercial CTL experience. The next chapter explains an acquisition process that the government could use to craft different packages to reflect differences in investor preferences.[2]

Given the complexities and ambiguities that arise when an investor uses debt financing, we focus here on robust incentive packages for equity financing. But, given appropriate assumptions about what debt share investors would choose and what probability of default to associate with each debt share at each average price of oil, we could use the same approach to choose robust incentive packages for debt financing. Given how powerfully debt share affects

---

[1] The equivalent range in pretax nominal terms is 13 to 27 percent. Note that this places the *average* prices for the entire period from the present to 2030 in the range of $40 to $70 per barrel. Since 1970, the mean world price of oil in 2006 dollars has been about $32 per barrel (median, $27 per barrel) despite the fact that prices approached $100 per barrel in present-day dollars in the early 1980s (WTRG Economics, undated). Although prices have once again reached and even passed $100 per barrel, no econometric model of the world oil market of which we are aware at this time points to average prices above $100 per barrel in present-day dollars over the next few decades. For anyone who doubts the models available today and expects average world prices to reach this level, our analysis makes it clear that first-of-a-kind CTL plants would be highly profitable without any special government treatment.

[2] The cash-flow model used here allows us to explore a wide range of alternative policy specifications rapidly, with little additional effort.

private IRR, those financial-incentive packages would in all likelihood be far less aggressive than those described next.

Packages suitable when project costs match case A differ from those suitable in case B.

Robust Financial-Incentive Packages for Case A

Table 6.1 displays information about four incentive packages that narrow and that realize real after-tax IRR outcomes when the investor uses 100-percent equity financing and case A holds for project and CO_2 management costs. Each column contains information about one incentive package. The null case includes no government incentives; packages 1 through 4 each include a different package of incentives. The first four rows show whether each package uses income sharing, what form of tax depreciation each uses, what level of investment tax credit each uses, and what CTL-based-fuel price floor each uses. Package 1 does not use income sharing; the others do. Packages 1 and 2 use standard DDB tax depreciation; packages 3 and 4 expense all investment costs in the first year of production. All apply a (nonbinding) fuel price floor of $40 per barrel.[3]

Figure 6.1 places these four packages in the IRR-NPV space used in Figure 4.1 in Chapter Four. Each package appears as a thread connecting (IRR, NPV) pairs for different average world oil prices, which are indicated by knots in each thread. Package 1 appears as a blue thread, package 2 a green thread, package 3 a red thread, and package 4 a yellow thread. The average world oil prices relevant to the knots in these threads are shown adjacent to the appropriate knots in the yellow thread representing package 4. The dotted black thread represents the null policy case, in which the government takes no positive action to encourage investment. The red dotted lines now demark the upper and lower boundaries of our target range for real after-tax IRR: 6 to 15 percent.

Note first in Figure 6.1 that, by design, each package achieves a fairly narrow range for private IRR by focusing increases in IRR on futures with lower oil prices. The range for government NPV, however, is much broader than in the null-policy case. The packages achieve robustness for real private after-tax IRR by increasing the range of outcomes for the government, effectively shifting risk to the government. This is potentially an appropriate strategy because, according to the aforementioned contracting principles, the government is better able to bear risk than most private investors are.

Table 6.1 displays this information numerically. Rows 5 and 6 show the real private after-tax IRR when the average price of oil is $40 and $70 per barrel, respectively. Rows 7 and 8 show the corresponding values of government NPV. Rows 9 and 10 show the change from the null case, at an average oil price of $40 per barrel, in investor IRR and government NPV, respectively. Row 11 shows the *average* CRIOP associated with moving from the null-policy case to each of the four policy cases shown. Rows 12 through 14 show results for an average oil price of $70 per barrel that are analogous to the results shown in rows 9 through 11.

---

[3] Keep in mind that the price floors shown here are *effective* price floors—the average price received by the investor when a price floor is binding. The legally stipulated price floors that yielded such prices received would lie somewhat below those identified in Table 6.1. How far depends on the volatility of future oil prices.

Table 6.1

Outcomes for Robust Financial-Incentive Packages When Case A Applies

Row	Measure	Null Case	Package			
			1	2	3	4
1	Income sharing?	No	No	Yes	Yes	Yes
2	Tax depreciation	DDB	DDB	DDB	100%	100%
3	Investment credit (%)	0	10	0	0	10
4	Price floor ($/barrel)	None	40	40	40	40
5	Investor IRR at $40/barrel (%)	5.96	6.66	5.96	6.30	7.39
6	Investor IRR at $70/barrel (%)	13.67	14.75	13.21	14.09	15.65
7	Government NPV at $40/barrel ($ millions)	−503	−622	−503	−705	−957
8	Government NPV at $70/barrel ($ millions)	897	777	1,021	790	539
9	Change in investor IRR at $40/barrel (%)		0.70	0	0.34	1.43
10	Change in government NPV at $40/barrel ($ millions)		−119	0	−202	−454
11	Average CRIOP at $40/barrel ($ millions)		170	NA	594	317
12	Change in investor IRR at $70/barrel (%)		1.08	−0.46	0.42	1.98
13	Change in government NPV at $70/barrel ($ millions)		−120	124	−107	−358
14	Average CRIOP at $70/barrel ($ millions)		111	270	255	181

The *average* CRIOP shown in Table 6.1 (and Table 6.2) is a rough measure of the relative effect of a financial-incentive package on the government and investor in any potential future. It differs from the marginal CRIOPs reported in previous chapters because it reflects the total effect of moving from the null-policy case to one of the policy cases shown here, not the incremental effect of adding one instrument at a time or changing the level of an instrument, such as a price floor or tax credit. As the discussion that follows will suggest, the ultimate choice among incentive packages must consider the marginal effects of each of their constituent elements in the presence of the others. By adjusting each element incrementally and observing the effect on CRIOP, a policymaker can choose a final incentive package by continuing adjustments until no further improvement is possible. We do not demonstrate that here but apply the logic underlying this approach to compare elements of the packages examined in this chapter.

The values of average CRIOP for package 1 dominate all others in Table 6.1 when a comparison is possible, suggesting that, if the government seeks a financial-incentive package likely to place private real IRR between 6 and 15 percent for average world oil prices of

Figure 6.1
Outcomes for Four Robust Financial-Incentive Packages When Case A Applies

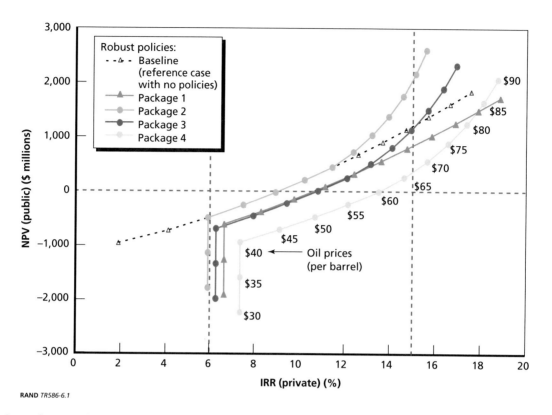

RAND *TR586-6.1*

$40 to $70 per barrel, then a package with no income sharing, DDB depreciation, a 10-percent investment tax credit, and a fuel price floor no higher than $40 per barrel is preferred. Package 1 dominates primarily because, as we observed in Chapter Four, an investment tax credit is such a cost-effective way to increase IRR relative to the alternatives. Policymakers might prefer the other packages to package 1 only if they were particularly concerned that values of IRR above 15 percent at average world oil prices above $70 per barrel were higher than necessary to achieve the objective at hand.

Looking beyond the preferred package, this framework also allows us to gain additional insights into how various packages compare. As one example, packages 3 and 4 allow us to see the effect of introducing a 10-percent investment tax credit, this time when 100-percent expensing prevails rather than DDB depreciation. Doing so again drops the average CRIOP at all prices, confirming the cost-effectiveness of the tax credit.

Comparing packages beyond those presented here would give policymakers many similar insights. Raising the oil price floor or investment tax credit above the levels shown here would narrow the range of private IRR more than our target range requires. Raising the investment tax credit could push IRR above 15 percent in our average world oil price range of interest, and package 4 already pushes IRR beyond our range at an average world oil price of $70 per barrel. Application of a finer mesh of values for oil price floors and investment tax incentives and tailoring the income-sharing agreement would allow policymakers to craft packages that matched the target range more closely in a more cost-effective way from the government's perspective. That said, these packages are likely to look, qualitatively, very much like the packages shown here.

Robust Financial-Incentive Packages for Case B

Table 6.2 displays information about the two packages that most closely match our robustness criteria when case B applies for project and CO_2 costs. It uses the same format as Table 6.1. Figure 6.2 graphically displays these two packages and compares them to the null case, in which the government takes no positive actions to encourage investment. Package 1 now appears as a blue thread, and package 2 as a green thread. Because the two packages differ only in their use of net-income sharing, these threads overlap below an average world oil price of $60 per barrel.

The range for private real after-tax IRR is, by design, about the same as that for packages chosen for case A. But because project and CO_2 costs are higher now, this can occur only if the NPV of cash flows to the government falls substantially. The large vertical distance between the null-case thread and the blue and green threads reflects this drop; a comparable vertical

Table 6.2
Outcomes for Robust Financial-Incentive Packages When Case B Applies

Row	Measure	Null Case	Package 1	Package 2
1	Income share?	No	No	Yes
2	Tax depreciation (%)	DDB	100	100
3	Investment credit (%)	0	25	25
4	Price floor ($/barrel)	None	45	45
5	Investor IRR at $40/barrel (%)	1.25	6.24	6.24
6	Investor IRR at $70/barrel (%)	9.75	14.15	14.06
7	Government NPV at $40/barrel ($ millions)	−1,257	−2,710	−2,710
8	Government NPV at $70/barrel ($ millions)	143	−897	−913
9	Change in investor IRR at $40/barrel (%)		4.99	4.99
10	Change in government NPV at $40/barrel ($ millions)		−1,453	−1,453
11	Average CRIOP at $40/barrel ($ millions)		291	291
12	Change in investor IRR at $70/barrel (%)		4.40	4.31
13	Change in government NPV at $70/barrel ($ millions)		−1,040	−1,056
14	Average CRIOP at $70/barrel ($ millions)		236	245

Figure 6.2
Outcomes for Robust Financial-Incentive Packages When Case B Applies

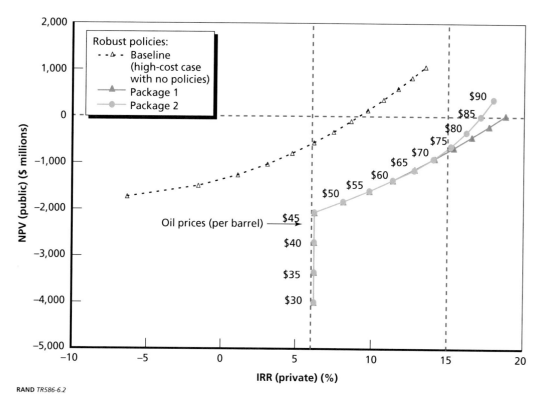

distance does not exist for any package considered in case A. For the robust packages we identified for case A, the real government NPV to implement them falls by about $0 to $450 million at various average oil prices, relative to the null-policy case. In case B, the real government NPV required to implement the robust policy cases identified falls much more—$1.0 billion to $1.5 billion at various world oil prices. For comparable circumstances, the packages in case B can cost the government $0 to $1.5 billion more than those in case A.[4] This potential difference may be large enough for policymakers to decide that a CTL production facility is too costly in case B. If they decide to proceed, government NPV will continue to vary widely across average oil prices within each case; robustness for private IRR occurs by shifting risk to the government.

The two robust incentive packages chosen to reflect circumstances in case B differ somewhat from those chosen for case A. Both packages expense investment costs immediately instead of using DDB tax depreciation. Both use the highest investment tax credit we considered to construct packages, 25 percent. Price floors for CTL-based fuel now bind at $45 per barrel to keep private IRR within the target range when average oil prices are low. The two case B packages differ only in their use of income sharing. One uses income sharing and the

---

[4]   This difference can be calculated in many ways. To get a rough assessment of the cost of maintaining an acceptable level of private after-tax IRR, we ranked the four robust packages we identified for case A by their real government NPVs at an average oil price of $40 per barrel, ranked the two we identified for case B in the same way, matched case A and B packages by rank, and took the difference for each match. We repeated this process at an average oil price of $70 per barrel. The range in the text reflects the range of all of the differences computed.

other does not; the difference is quite small, because the sharing arrangement we used has little effect at $70 per barrel. In sum, the structure of a balanced incentive package changes subtly when expectations about project costs change.

As in case A, application of a finer mesh of values for oil price floors and investment tax credits and tailoring the income-sharing agreement would allow policymakers to craft packages that matched the target range more closely in a more cost-effective way from the government's perspective. That said, such packages are likely to look, qualitatively, very much like the packages do here.

Ideally, we would have sought financial-incentive packages that allowed higher levels of private real IRR at higher prices, but the range of policy parameters we considered would not allow that. The options for allowing higher levels of private real IRR at higher prices are limited. Higher price floors would have no effect, and no further enhancement of tax depreciation is possible. Removing income sharing has only a minor effect at $70 per barrel. That leaves a more aggressive investment tax credit. Twenty-five percent is quite high already. But, as we noted in Chapter Four, one of the key advantages of the investment tax credit is that increases in it can enhance private real IRR over wide ranges of values. In principle, we could use a tax credit to push private real IRR as high as required.

Of course, debt financing would also push real private IRR higher. An analysis that reflected the effects of such financing would yield less aggressive incentive packages than those described in this chapter. This framework can be used to address future debt financing if a reasonable basis can be established for choosing debt shares likely to prevail over the life of a project facing different average oil prices.

CHAPTER SEVEN

Can Formal Source Selection Help the Government Create an Integrated Policy?

The discussion in Chapter Two (quite deliberately) examines the policy instruments that the government might use to induce private participation in a project in terms of language normally used to discuss the design of voluntary agreements. The principles in Chapter Two spring from historical experience with voluntary agreements that have successfully survived the test of time in competitive environments. These principles promote the use of various policy instruments in designing voluntary agreements. Each agreement uses a different set of instruments to reflect the specific characteristics and mutual interests of the parties to the agreement. Agreements between the government and potential investors might be created in a similar way. How could this be done?

The government could hold a formal source selection that asked potential offerors to come forward with (1) specific technical plans to build and operate a specified plant and monitor information about it, (2) specific management plans to ensure financial capacity and effective control over these activities, (3) specific policy instruments that they would want applied to manage their relationship with the government over the project's life, and (4) historical evidence that they can execute the plans described in the first three points. The third element is where specific mutual agreements could be crafted. Mechanisms exist for the government to discuss the structure of alternative agreements with individual offerors as long as the discussions do not privilege any one offeror and they protect proprietary information from offerors.[1] These mechanisms must be designed with great care to avoid abuse that could support successful appeals of source-selection decisions. But such mechanisms are available.

The source-selection authority would then compare best and final offers against a set of clearly stated standards and choose the offers that match those standards most closely. To make this approach work, the government must be able to state these standards in a way that can stand up to appeal. The standards must be clear enough that reasonable, objective observers can agree on how closely any offer matches each standard. In this setting, the government would have to be able to state what effects it wanted any set of policy instruments to achieve. Three key elements would likely be important:

- likelihood that the investor would, in fact, successfully build a plant
- likelihood that the investor would successfully achieve substantial early CTL production experience

---

[1] For some examples of how this has been done in recent federal acquisitions, see Camm, Blickstein, and Venzor, 2004. The *Federal Acquisition Regulation* (GSA, DoD, and NASA, 2008) provides the detail on how to structure individual source selections to comply with federal law and policy.

- given the first two, the expected cost to the government of implementing the policy instruments specified in an offer.

In all likelihood, the government would have to develop a formal framework for addressing each of these issues and share the framework with potential offerors early in the source selection. The government could even offer the framework in draft form, take input from potential offerors, and adjust the framework before calling for formal offers. Such an approach would allow the government to encourage dialogue with interested parties and develop a set of policies that best reflects the ideas these parties have to offer.

Why design financial-incentive packages in a source selection? Why not use an administrative process that does all of these things and then let the government select the policy instruments it will use to manage its relationships with investors? Such a process would impose less demanding requirements for clarity and specificity about many issues that are inherently subjective and strongly colored by uncertainty about the future. This is an option that should be explored. But if, in the course of such an approach, the government concluded that it could state its priorities clearly enough to use a formal source selection to choose packages of policy instruments, a formal source selection could

- create incentives that encouraged offerors to be as honest and accurate as possible about the basis for the proposals they bring forward (each wants to win, but none wants to win if it cannot execute what it promises)
- collect this information in a way that allows investors with very different preferences among alternative incentive packages to participate and allows the government to engage all of them on a level playing field
- given information generated in this way, allow the government to craft different sets of policies for different investors to reflect their priorities and take advantage of these to the mutual interest of government and investor.

In such an approach, the qualitative principles discussed in Chapter Two and the policy instruments that are compatible with them would likely play a central role. A cash-flow analysis of the kind described in Chapters Three through Six would be critical to any objective government assessment of the likelihood of achieving early CTL production experience and the cost to the government of making that happen. But even if the government used a less formal, more traditional administrative process to choose the policy instruments it uses to promote private participation, the considerations discussed here would still play a role. In particular, they should help the government elicit and integrate useful information from potential offerors as it worked out what set of policies it wanted to implement.

CHAPTER EIGHT

Conclusions

This technical report is based on the assumption that the government, as a principal, seeks to induce a private investor, as an agent, to build and operate an unconventional-oil production plant to promote early production experience with such plants. Given this goal, facing significant uncertainty about the future, the government wants to limit the cost to the public treasury of doing this. This report offers an analytic way to design a package of policy instruments that the government can use to achieve its goal.

It starts with general principles of the economic theories of contracting and agency. These remind us that the structure of incentives subtly affects what kinds of investors are attracted to such a project and how cost-effectively they will build and operate an unconventional-oil production plant. Up to a point, the more risk the government can shift to an investor, the better the performance the government can expect from the investor. But this theory also reminds us that the government is better able to bear risk than most investors are and has a lower cost of capital. Opportunities exist to exploit these differences between government and investor to their mutual advantage. These principles tell us that it is worthwhile, if done properly, to pursue robust policies that limit the range of expected return on private investment by shifting risk to the government, increasing the range of likely costs to the government. They also tell us to expect incentives to be more cost-effective when the government aids the private investor as early in a project as possible. Following such principles will shape the cash flows from any project in positive but generally quantitatively unpredictable ways.

Given any set of expected cash flows, quantitative spreadsheet analysis allows detailed assessment of the effects of different financial-incentive packages on the private rate of return from a project and the project's cost to the government when different versions of the future transpire. Looking across many alternative futures allows us to design incentive packages that are robust from a private perspective and that limit costs to the government. As these principles would predict, cash-flow analysis demonstrates the cost-effectiveness of using investment incentives rather than operating incentives and the powerful effect that a higher debt share has on the private rate of return. Cash-flow analysis also reveals specific opportunities that the government has to change course among policy alternatives as it seeks the lowest-cost way to increase the private rate of return associated with a project.

Such analysis reveals the desirability of a balanced package of incentives that uses a floor on the oil price that an investor receives when the average oil price is low, investment tax credits to improve the private rate of return on investment at higher average oil prices, and income-sharing agreements (much like those used in petroleum production contracts around the world today) that share net revenue from a plant between the investor and government when average oil prices are high.

Such analysis also reveals the power of loan guarantees to encourage private investment. But the incentive effects of such guarantees flow from an implicit expectation that the government will exercise its responsibility to pay off a portion of the loan. These effects are especially insidious because, unless government policy stops it, the presence of a loan guarantee can encourage investors to pursue a dangerously high debt share. In effect, the higher the debt share, the higher the likelihood that the government will have to pay a portion of the loan, and the lower the investor expects the cost of the loan to be. Combined with a tendency to draw less satisfactory investors to a project, these effects of loan guarantees demand that the government use guarantees only with great care and a full appreciation of how much government oversight is required to limit deleterious investor behavior.

Structure of the Spreadsheet Analysis That Implements the Cash-Flow Model

Following standard financial practice, we constructed a project cash-flow model to calculate (1) the real (adjusted for inflation) NPV and IRR of cash flows for a company investing in a combined-cycle CTL production plant and (2) the real NPV and IRR of cash flows that the government associates with the plant. Cash flows relevant to the government include taxes from the plant and subsidies to the plant over its lifetime. Key factors that the model allows to vary are (1) the benchmark price of crude oil, (2) the cost of construction and maintenance, and (3) various government incentives designed to promote early private-sector CTL production experience.

This appendix explains the basis for the input values assumed in the model. It then explains the structure of the cash-flow model itself. Calculations use end-of-2006 dollars throughout.

Input Values Assumed in the Model

Basic Plant Parameters

SSEB (2006, Appendix D) presented an FT CTL plant, which it called case 3, that consumes 17,987 tons per day of bituminous coal feedstock to produce approximately 30,000 barrels per day of liquid fuels. Unless otherwise noted, we drew the values of all parameters used in this appendix from that report.[1]

The plant's daily output is calculated on a DVE basis. In fact, the volumetric output of the plant would comprise 24,359 barrels of FT diesel and 11,398 barrels of FT naphtha, a sum of 35,757 barrels. The naphtha is assumed to have a value that is 71 percent of that of the diesel product. In terms of DVE, the plant output is 32,502 barrels per day (SSEB, 2006, Appendix C, p. 7).

The CTL plant also generates a total of 725 megawatts (MW) of electric power. Of this, 521 MW would be used to operate the plant itself, and the remaining 204 MW would be sold to the electricity grid.

In addition to producing synthetic fuel and electricity, the plant produces 24,734 tons of CO_2 per day. Table A.1 summarizes these assumed characteristics of the plant.

Plant Construction Expenses

We consider two alternative assumptions about costs: a reference case and a high-cost case. In the reference case, the total capital cost of the plant is $3.305 billion. The number breaks down

---

[1]   Bartis, Camm, and Ortiz, 2008, Appendix A, discusses in greater depth the basis for the assumptions used here.

Table A.1
Summary of Plant Intake and Production

Measure	Capacity
Plant input: coal intake (tons/day)	17,987
Plant outputs	
Diesel fuel (barrels/day)	32,502
Electricity (MW)	204
CO_2 (tons/day)	24,734

as follows: The total cost for capital equipment, according to the SSEB report, is $2.224 billion. We added a 25-percent contingency for cost growth to reflect the first-of-a-kind nature of the plant. We also added 6.56-percent growth in the cost of refinery construction to update the SSEB (2006) numbers to the end of 2006 (Farrar, 2007). Consistent with the SSEB report, we assume $343 million in home-office expenses, start-up costs, and working capital.

Of these capital costs, we assume that all but the $343 million in start-up and working capital is spent over a five-year period of construction. When the plant begins operation the following year (operating year 1), start-up costs and working capital costs are incurred. We assume that start-up and working capital costs are expensed and that capital costs are depreciable on a Modified Accelerated Cost Recovery System (MACRS) seven-year DDB schedule.

In runs of the model using the high-cost case, we assume that all capital expenses except the start-up and working capital increase by an additional 25 percent. Table A.2 summarizes these assumed characteristics of the plant.

Plant Operational Expenses

We assume that the plant operates at 70 percent of capacity during its first year of operations and at 90 percent for all remaining years. In the reference case, we assume that the fixed annual operating costs of the plant are $131.8 million, which covers labor and labor overhead, administrative, local taxes and insurance, and maintenance and material costs (SSEB, 2006, Appendix D, page 39). The variable noncoal operating costs are $2.69 per barrel, based on the SSEB assumption that royalties and catalyst chemicals would cost $28.7 million in a full-production year. The variable fuel costs of $16.50 per barrel are based on a coal requirement of 0.55 tons per barrel of product (DVE) and a coal price of $30 per ton. At an annual capacity factor of 90 percent, total operating costs are $336.7 million per year.

Table A.2
Summary of Plant Capital Expenses

Capital Expenditure	Amount ($ millions)	Depreciation	In High-Cost Case
Capital equipment	2,224	7-year DDB	Increases by 25%
25% contingency	556	7-year DDB	Increases by 25%
6.56% cost growth	182	7-year DDB	Increases by 25%
Start-up and working capital	343	In year 1	Does not change

In the high-cost case, we increase the fixed operating costs by 33 percent while leaving the variable costs the same. Table A.3 summarizes these assumed plant characteristics.

Income

The plant's total income is generated from the production and sale of the three products specified in Table A.1: diesel fuel, electricity, and CO_2.

We assume, based on historical data, that the sale price of diesel fuel is 1.3 times the benchmark price of crude oil. This factor does not change as we vary the price of crude oil in the analysis. We assume a sale price of $0.050 per kilowatt-hour (kWh) for grid electricity and hold this price constant throughout the analysis. We assume that coal costs $30 per ton. We considered scenarios in which the CO_2 is sold for enhanced oil recovery (EOR) or costs $10 per ton to dispose of, but we generally assume that disposal of excess CO_2 incurs no additional costs.

Cash Flows

Cash Flows for the Company, Assuming No Project-Specific Public Policies

We followed standard financial practice in constructing our cash-flow model. During the construction period, net after-tax cash flow comprises only the level of construction outlays made in a given year. During the 30 years of assumed plant operation, we calculate gross income as the difference between net income and net operating costs. From this figure, we subtract depreciation and expenses (including start-up and working capital) before calculating taxes. We assume throughout that the company can take full advantage of depreciation charges and tax benefits in the year in which they become available. We assume a federal tax rate of 34 percent and a state tax rate of 3.2 percent.

When the company uses debt financing, we assume that it borrows incrementally in each year that it invests and pays interest on any debt incurred until the principal is repaid. We schedule the repayment of principal to generate a constant real annual cost of servicing the loan over the production phase of the project. Annual payments are almost entirely interest repayments at the beginning of production and almost entirely principal repayments at the end of the production phase. We deduct interest payments each year from taxable income.

We calculate the real after-tax IRR to the company using an iterative procedure to determine the discount rate that sets real NPV to 0. Figure A.1 summarizes these calculations.

Table A.3
Summary of Plant Operating Expenses

Operating Expense	Amount (per barrel)	Year 1	Years 2+	In High-Cost Case
Fixed operating costs		$131.8 million	$131.8 million	Increases by 33%
Plant availability (%)		70	90	
Variable plant costs	$2.69	$22.3 million	$28.7 million	No change
Variable fuel costs	$16.50	$137.0 million	$176.2 million	No change
Total		$291.1 million	$336.7 million	

Figure A.1
High-Level Structure of Cash
Flows

```
                                    revenue
                         – operating costs
                            ───────────────
                              gross income
                            – depreciation
                               – expensing
                            ───────────────
                            taxable income
                                   – taxes
                            ───────────────
                                net income
                            + depreciation
                               + expensing
                          – capital expenses
                            ───────────────
                 net after-tax cash flow
```

RAND *TR586-A.1*

Cash Flows for the Government, Assuming No Project-Specific Public Policies

We calculate the net value of cash flows to the government in much the same way, taking the difference between income and expenditures in a given year, as summarized in Figure A.2. If the company uses debt financing, we treat interest payments as tax deductible for the company but as taxable income for its lender. We assume that the lender faces the same standard tax structure as the company does. We apply a discount rate to all project-associated cash flows to and from the government over the life of the project to calculate an NPV of the project to the government.[2] As noted in the text, we use the real discount rate prescribed by the Office of Management and Budget (OMB), 7 percent, for government programs of the type examined here (see Chapter Three of this report).

Figure A.2
Treatment of Taxes

```
                        tax revenue collected
       – deductions for depreciation of capital
         – deductions for expensing of capital
          ─────────────────────────────────
                net government cash flow
```

RAND *TR586-A.2*

---

[2] The model also allows calculation of the real internal rate of return to the government by using an iterative procedure to determine the discount rate that sets real net present value of cash flows to the government equal to zero. We can calculate this rate, even in the null policy case, where no policy incentives are in place for this project, because the government still gains and loses money from the project, due to tax revenue collected and tax deductions allowed (such as depreciation). We did not use this value of real IRR for the government in the analysis reported in this document.

There is no scenario in which the government is completely unaffected by the plant's construction. Even if the project were abandoned following construction, it would represent a net loss for the government, as the depreciation would be taken without having generated tax revenue.

The Financial Effects of Public Policies

Up to this point, discussion has focused on the public and private cash flows that result when no government policy incentives are in place to promote production. The six policy incentives discussed in this report typically involve a cost incurred by the government in exchange for a benefit delivered to the company. Income sharing is the exception. Figure A.3 summarizes the five types of effects that these policies can have and that the cash-flow model captures.

The six policy incentives have specific effects, in terms of their impact on cash-flow streams. In the case of purchase guarantees, price guarantees, and price floors, government funds are infused directly into the company's income stream. In contrast, direct investment and operational credits seek to reduce the company's costs. All four of these policy effects, whether they guarantee income or reduce costs, assist the company by making its pretax gross income higher.

A loan guarantee has quite a different effect. As discussed in Chapter Two, a loan guarantee transfers the risk of loan default from a third-party lender to the government, thereby reducing the cost of the loan (i.e., the interest rate) to its recipient. The government accepts as a cost the probability that the company could default—and therefore, the probability of default multiplied by the loan amount constitutes the cost of any given loan guarantee. Figure A.4 shows how the cash-flow model captures these effects.

Guaranteed Income Streams

As mentioned in the previous section, the effect that purchase guarantees, price guarantees, and price floors have on the private firm is to supplement private income with a direct expenditure by the government.

Figure A.3
The Five Effects Our Policy Instruments Have on Cash Flows

Policy instruments	Effect
Purchase or price guarantees Price floor	Company receives guaranteed income
Investment credits Direct subsidy Operational credits	Company receives lower costs
Investment credits Tax subsidy	Company receives greater tax deductions
Income-sharing arrangement	Government receives more tax revenue
Loan guarantees	Company receives lower cost of debt

RAND TR586-A.3

Figure A.4
Effect of Direct Subsidies on Income Streams

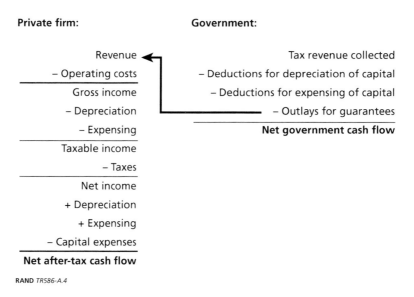

RAND *TR586-A.4*

In Figure A.4, a new entry in the government column shows a flow of spending that goes directly from the government to the private firm for such guarantees. It should be noted, however, that the amount of income that the company receives is not always equivalent to that paid out by the government. In the case of a purchase guarantee at market price, the cost to the government is actually zero, because we assume that the government would have needed to purchase fuel anyway and would have paid market price for it. In the cases of the price guarantee and price floor, the government's cost is actually the difference between the guaranteed price and the market price when the guaranteed price is lower. If the guaranteed price is higher, the price floor simply reverts back to market price, whereas, in the price guarantee, the government stands to gain money.

Cost-Lowering Policies

The effect that direct investment and operational credits has on the private firm is to reduce the private firm's costs of construction and operation. In both cases, the government pays for these credits directly. Figure A.5 shows how the cash-flow model captures these effects.

In our model, we assume that all construction expenditures take place during a five-year period and that operation begins the following year. At no time are both policies affecting costs in the same year. Outlays for cost subsidies constitute a new entry in the government column because they represent a direct transfer of funds. We should also note, however, that, in the case of capital investment subsidies, the amount of depreciable capital is reduced by the amount of subsidy received, and the government thus recovers a portion of the cost of its subsidy through slightly higher taxes.

Greater Tax Deductions

The other type of investment credit discussed in this report is the use of tax deductions to lower the amount of taxes the private firm pays over the project's life. Figure A.6 shows how the cash-flow model reflects these.

Figure A.5
Effect of Direct Subsidies on Cost Streams

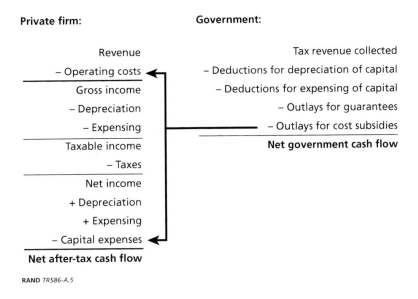

Figure A.6
Effect of Tax Deductions

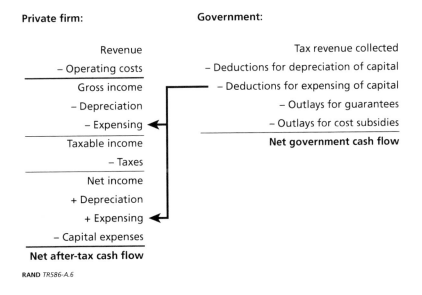

Income-Sharing Arrangements

This policy incentive differs from the rest of those discussed here because the government stands to benefit at the company's expense. In the examples used in our analysis, the government could demand such an arrangement when the benchmark crude oil price is high (above $60 in our analysis), in exchange for incentive packages extended to the company at lower oil prices.

In our analysis, we simulate income sharing by employing higher tax rates as oil prices become higher. Whereas the default tax rate is 36.1 percent (a combination of federal and state rates), the taxes rate rises to 48.1 percent when the average oil price reaches $70 per barrel, 60.0 percent when the average oil price reaches $80 per barrel, and 68.1 percent when the average oil price reaches $90 per barrel. Figure A.7 shows how the cash-flow model captures the effect of such an income-sharing agreement.

Loan Guarantees

As discussed earlier, a loan guarantee transfers the risk of loan default to the government and reduces the recipient's cost for the loan. The government accepts as a cost the probability that the company could default. Therefore, the probability of default multiplied by the loan amount constitutes the cost of any given loan guarantee.

It is difficult to illustrate the effect of a loan guarantee in the same way that we have for the other four policy effects because it does not represent a direct transfer of funds from the government to the private firm. Both the private company and the government deal directly with the lender. However, it is still possible to demonstrate where both of these transfers fit into the cash flow. Figure A.8 illustrates this.

The cash-flow diagram is the same for the private firm as it was before—the firm continues to pay back its consolidated loan during operating years, counting its principal repayment as part of its capital expenditures. On the other hand, a new entry is shown in the government column, denoting the outlays it will have to make to repay the loans it has guaranteed on which the borrower has defaulted.

Figure A.7
Effects of Income Sharing Through Higher Taxes

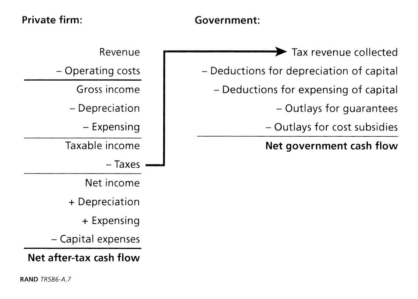

RAND TR586-A.7

Figure A.8
Cash-Flow Schematic Including Loan Guarantees

Private firm:	Government:
Revenue	Tax revenue collected
− Operating costs	− Deductions for depreciation of capital
Gross income	− Deductions for expensing of capital
− Depreciation	− Outlays for guarantees
− Expensing	− Outlays for cost subsidies
Taxable income	− Outlays for loan repayment
− Taxes	**Net government cash flow**
Net income	
+ Depreciation	
+ Expensing	
− Capital expenses	
Net after-tax cash flow	

RAND *TR586-A.8*

How Debt and Loan Guarantees Affect Investors and the Government

In principle, lenders present borrowers with a set of options for loan rates and debt share—loan rates rise with rising debt share—and borrowers choose a debt share that allows them to balance the value of access to capital that costs less than equity with the costs associated with default, which directly affect their expected future cash flows as well as the cost of debt capital available to them. Presumably, lenders condition the set of options available on the nature of the investment to be financed and the character of the investor seeking the loan.

This appendix seeks to elucidate the key factors that affect this relationship between borrower and lender when debt financing is available and then when it is available with a government-underwritten loan guarantee attached. It first presents a simple mathematical model of the key factors that drive the lender's and borrower's decisions. It then introduces a loan guarantee into this model and traces its key effects on those decisions and resultant cash flows to the lender, borrower, and government. It demonstrates clearly that all of the factors that make a loan guarantee attractive to lenders and borrowers stem from shifting the costs of default from themselves to the government. If they did not expect default, the presence of a government-provided loan guarantee would have no effect on their behavior. That is, any government assessment of the benefits of a loan guarantee must address the fact that the government finances all of these benefits directly from its own pocket.

Table B.1 summarizes the expressions that together comprise the model. Column 1 shows a series of expressions relevant to a situation in which an investor takes advantage of debt capital without a loan guarantee; column 2 shows analogous expressions with a loan guarantee. Consider the case without a loan guarantee first.

Effects of Debt Without a Loan Guarantee

Row 1 shows the value of a loan, the product of total investment cost, I, and the debt share of this total investment, s. When a lender makes a loan, it commits a total amount of monitoring resources, m_{LO}, per dollar of loan. From a lender's perspective, $sI(1+m_{LO})$ is the amount that it must recover from a borrower if the borrower does not default (row 2). Default in itself imposes additional costs on the lender, $c_{LO}sI$, which the lender will anticipate and plan to recover from a borrower to ensure that it recovers all its own costs of offering the loan (row 3). The borrower, lender, and government foresee the same probability of default, p.

If a default occurs, the lender expects to receive nothing from the borrower (row 4). In effect, in this model, the probability of default measures the share of the scheduled payments

Table B.1
Summary of Effects of Debt and Loan Guarantees on Investors and the Government

Row	Effect	1: Without Loan Guarantee	2: With Loan Guarantee
1	Loan value	sl	sl
2	Total opportunity cost to lender without default	$sl(1 + m_{LO})$	$sl(1 + m_{LG})$
3	Total opportunity cost to lender with default	$sl(1 + m_{LO} + pc_{LO})$	$sl(1 + m_{LG} + pc_{LG})$
4	Total payment lender receives in default	0	$sl(1 + m_{LG} + pc_{LG})$
5	Annual payment lender requires without default to cover opportunity cost	$\dfrac{D_L sl\left(1 + m_{LO} + pc_{LO}\right)}{1 - p}$	$D_L sl(1 + m_{LG} + pc_{LG})$
6	Annual payment that lender expects	$D_L sl(1 + m_{LO} + pc_{LO})$	$D_L sl(1 + m_{LG} + pc_{LG})$
7	Annual income for borrower without default	$N_B - \dfrac{Dsl\left(1 + m_{LO} + pc_{LO}\right)}{1 - p}$	$N_B - D_L sl(1 + m_{LG} + pc_{LG})$
8	Annual income for borrower with default	$-D_B c_B sl$	$-D_B c_B sl$
9	Annual income that borrower expects	$(1-p)N_B - [D_L(1 + m_{LO} + pc_{LO}) + pD_B c_B]sl$	$(1-p)[N_B - D_L(1 + m_{LG} + pc_{LG})]sl - pD_B c_B sl$
10	Annual direct payment by government without default	0	$D_G m_{GG} sl$
11	Annual direct payment by government with default	0	$[(1 + m_{LG} + pc_{LG})D_L + c_G D_G + m_{GG} D_G]sl$
12	Annual direct payment that government expects	0	$[(1 + m_{LG} + pc_{LG})pD_L + pc_G D_G + m_{GG} D_G]sl$

that the lender expects it will not receive. For a loan over n years, if the lender's cost of capital is r_L, the lender must receive D_L per year per dollar of loan to recover its cost of capital, where[1]

$$D_L \equiv \frac{1 - \delta}{\delta\left(1 - \delta^n\right)},$$

$$\text{where } \delta = \frac{1}{1 + r_L}.$$

(B.1)

As a result, when it does receive a payment, the lender must receive the annual payment shown in row 5 to ensure that it covers all expected costs of the loan. The cost of capital, r_L, reflects both the cost of funds that the lender commits to a loan and the lender's normal administrative costs beyond the costs of monitoring the loan itself. Per dollar of loan, the lender charges the borrower

---

[1]   This is based on a standard formula used by bankers to identify loan payments.

$$D_L \frac{1 + m_{LO} + pc_{LO}}{1-p}.$$

The lender presumably chooses a level of monitoring that cost-effectively limits the costs of default associated with p and c_{LO}. The resulting annual payment exceeds that required by the lender's basic cost of capital for three reasons. First, the factor $(1 - p)$ compensates for the basic probability of default. Second, the factor pc_{LO} compensates the cost of default itself, $c_{LO}sI$, which occurs with probability p. Third, the factor m_{LO} compensates for the lender's costs of monitoring the loan. Given that the lender expects to receive this amount with a probability of only $(1 - p)$, in this example, the lender expects to receive each year $D_L(1 + m_{LO} + pc_{LO})$ per dollar lent (row 6).

Given the amount shown in row 5, the borrower expects to earn the amount shown in row 7 when it does not default, where N_B is the annual net income the borrower expects prior to paying off its loan. When it does default, it expects no net income and does not expect to make any payment to the lender. But it does expect costs when it defaults, which may take the form of immediate administrative costs and longer-term effects on its access to credit. Row 8 represents this cost as proportional to the size of the loan, $D_B c_B SI$, where D_B is a discount factor analogous to that in Equation B.1 that the borrower uses to move between annual and total flows.

Row 9 shows the annual amount the borrower expects to earn when it uses debt capital. In effect, given that the borrower invests $(1 - s)1$ of its own equity in the project, the amount in row 9 must equal $(1 - s)ID_B$, where we can now understand that the r_B implicit in D_B is the borrower's IRR from the investment. To allow us to focus on the central structure involved, assume that $m_{LO} = c_{LO} = c_B = 0$. Then

$$\left(1 - s\right)ID_B = \left(1 - p\right)N_B - sID_L, \text{ or}$$

$$D_T = \frac{\left(1 - p\right)N_B}{I} = sD_L + \left(1 - s\right)D_B, \tag{B.2}$$

where D_T is the discount factor associated with the cash flows in the project before any loan payments. Rearranged, Equation B.2 becomes

$$D_B = \left(\frac{1}{1-s}\right)D_T - \left(\frac{s}{1-s}\right)D_L. \tag{B.3}$$

D_B, and therefore the borrower's IRR, rises as D_T rises or D_L (that is, r_L) falls. How does the debt share affect the borrower's IRR?

$$\frac{\partial D_B}{\partial s} = \left(\frac{1}{1-s}\right)\frac{\partial D_T}{\partial s} + \frac{D_T - D_L}{\left(1-s\right)^2}$$

$$= \left[-\left(\frac{N_B}{I}\right)\frac{\partial p}{\partial s} + \frac{D_T - D_L}{1-s}\right]\left(\frac{1}{1-s}\right).$$

(B.4)

If the debt share has no effect on the probability of default, debt share unambiguously increases D_B and therefore borrower IRR as long as $D_T > D_L$. If debt share increases the probability of default, as is likely at some point, an increase in debt share reduces the value of the net income the borrower expects from the project. The net effect on borrower IRR depends on which effect is larger at any level of probability of default as it increases:

$$\frac{\partial D_B}{\partial s} > \text{or} < 0 \text{ as } \left(\frac{N_B}{I}\right)\frac{\partial p}{\partial s} < \text{or} > \frac{D_T - D_L}{1-s}.$$

(B.5)

If rising debt share has no effect on the probability of default,

$$\frac{D_T - D_L}{1-s}$$

rises with rising debt share as $(1-s)$ rises. At some point, as debt share continues to rise, it will affect the probability of default. At that point, D_T will begin to fall as debt share continues to rise, reducing the slope of

$$\frac{D_T - D_L}{1-s}.$$

Figure B.1 displays what this might look like. Meanwhile, as rising debt share increases the probability of default,

$$\frac{N_B}{I}\left(\frac{\partial p}{\partial s}\right)$$

will rise from the horizontal axis. To maximize its IRR, a borrower chooses a debt share for which

$$\left(1-s\right)\left(\frac{N_B}{I}\right)\left(\frac{\partial p}{\partial s}\right)$$

just equals $(D_T - D_L)$, at s_0 in Figure B.1.

When m_{LO}, c_{LO}, and c_B are nonzero, the expressions get much more complicated, but the qualitative results are similar. Now

$$(1-s)ID_B = (1-p)N_B - D_L(1+m_{LO}+pc_{LO})sI + pD_Bc_BsI \text{ and}$$

$$D_B = \frac{D_T}{1-s-pc_Bs} - \frac{D_L(1+m_{LO}+pc_{LO})s}{1-s-pc_Bs}.$$

(B.6)

Figure B.1
How a Borrower Chooses a Debt Share to Minimize Cost of Capital

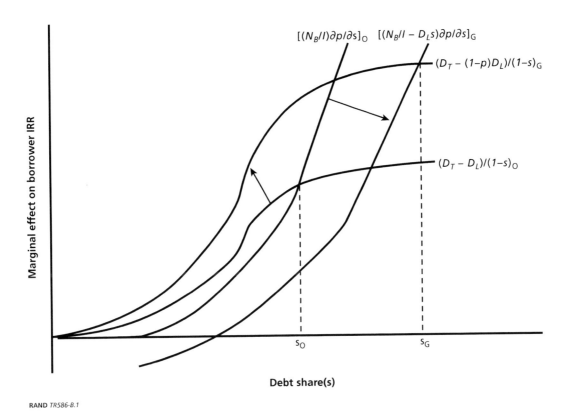

RAND TR586-B.1

In effect, the definitions of annual payments relevant to the lender and borrower have become complex, but the structure of Equation B.6 is the same as that for Equation B.3. Now, not only D_T, but also p, m_{LO}, c_{LO}, and c_B are functions of debt share. These additional factors tend to have smaller effects than the ones examined before. Including them leads to a lower optimal debt share simply because each of them involves costs associated with using debt financing, and each of these costs tends to rise as the debt share rises. Otherwise, including them does not change the qualitative findings presented here or the intuition underlying them.

The government has an interest in this loan only in terms of how it might affect the tax revenue it collects. This simple model sets tax revenue effects aside, leaving the government unaffected by the loan.

Effects of a Loan Guarantee

Now consider the same series of considerations when the government guarantees the full value of the loan in question. Column 2 of Table B.2 shows this case. Row 1 is the same as before. Row 2 has changed to reflect the likelihood that a lender that is now guaranteed payment on its loan will reduce its own monitoring of the loan. The lender's monitoring cost per dollar of loan with a loan guarantee, m_{LG}, is smaller than m_{LO}, its monitoring costs without a loan guarantee. Similarly, row 3 has changed to reflect the possibility that the lender may now face a smaller—and potentially zero—cost if a default occurs. That is, c_{LG} is probably smaller than c_{LO} and potentially zero.

With a loan guarantee, even if a default occurs, the lender receives the full amount required to cover its costs of offering the loan (row 4). As a result, the lender now requires the amount shown in row 5 from a borrower as an annual payment, revealing that the lender (row 6) expects to receive $D_L(1 + m_{LG} + pc_{LG})sI$, which is lower than the equivalent cost without a loan guarantee when m_{LG} is lower than m_{LO} or c_{LG} is lower than c_{LO}.

Because the lender is paid with or without a default, when the borrower does not default, the borrower can now expect annual income of $N_B - D_L sI(1 + m_{LG} + pc_{LG})$ (row 7). This is higher than in the absence of a loan guarantee for three reasons. First, the payment need not cover the possibility of default because the government now does this. Second, the lender faces smaller costs following default and so need not recover as much from the borrower to cover these costs. Third, the lender faces smaller monitoring costs for the loan and so need not recover as much from the borrower to cover these costs. When the borrower defaults, it suffers the same loss it would without a loan guarantee (row 8). Row 9 shows the annual amount of income the borrower expects to earn when it uses debt capital. To the extent that any risk of default exists, this is higher than without a loan guarantee for all three reasons discussed with regard to row 7. That is, the borrower expects the amount it will pay per dollar of a loan to fall from

$$\left[D_L\left(1 + m_{LO} + pc_{LO}\right) + pc_B \right] - \left[\left(1 - p\right)\left(1 + m_{LG} + pc_{LG}\right) + pc_B\right]$$

(row 9). This reduction will increase the borrower's IRR.

Note that *all* the factors that lower the cost of debt capital for the borrower when a loan guarantee is available reflect elements of default. The borrower no longer has to pay an interest rate designed to cover the lender's full normal costs of business if default occurs because someone else—the government—now does this. The borrower probably does not have to pay as much to cover the direct costs of default to the lender, $c_{LG}sI$, because those costs are now lower. Again, someone other than the lender—the government—bears analogous costs without asking the borrower to reimburse these costs. In sum, the way a loan guarantee reduces a borrower's cost of debt capital is to shift costs associated with default from the borrower to the government.

The government now has a direct interest in the loan, over and above its interest in tax revenue. Even in the absence of default, the government now has a fiduciary responsibility to monitor the loan. This imposes a cost, $m_{GG}sI$, over the life of the loan (row 10). The level of this cost presumably reflects a government decision to conduct monitoring at a level that discourages default as long as the cost of the monitoring required to do this does not exceed the savings from default costs that monitoring produces. The government faces a constant trade-off between accepting monitoring costs and accepting default costs. If the borrower defaults, the government faces an immediate liability implied by the payment identified in row 5. Evaluated at the discount factor compatible with the government's cost of capital, D_G, the annual value of this liability is

$$\left[\left(1 + m_{LG} + pc_L\right)D_L + c_G D_G + m_{GG}D_G\right]sI$$

(row 11). The monitoring costs discussed in row 10 occur here as well. And the government has its own costs if a default occurs, $c_G sI$, which may be primarily administrative or may reflect a default's effect on the government's ability to use loan guarantees in the future.

The government's expected annual cost of offering a loan guarantee, if it shares lender and borrower expectations about default, as assumed here, is

$$\left\{p\left[\left(1 + m_{LG} + pc_L\right)D_L + c_G D_G + m_{GG}D_G\right]\right\}sI$$

(row 12). Government costs rise most directly when the probability of default or the cost of default to the lender or government rise. They rise when monitoring costs rise, to the lender or the government, but either form of monitoring presumably reduces the probability of default and perhaps its administrative costs to the government and its effects of future government freedom to use guarantees again.

Given that a loan guarantee benefits a borrower by shifting costs associated with default from the borrower to the government, we should not be surprised if the borrower made a decision under a loan guarantee that led to a higher default rate than it would choose without a loan guarantee. This simple model allows us to see how this occurs. Again, to focus in the central features of the model, assume that $m_{LO} = c_{LO} = c_B = 0$. Then, with a loan guarantee, the borrower's view of its expected annual income from the project becomes

$$\left(1 - s\right)ID_B = \left(1 - p\right)N_B - \left(1 - p\right)sID_L, \text{ or}$$

$$D_B = \left[\left(\frac{1}{1-s}\right)D_T - \left(\frac{s}{1-s}\right)\left(1 - p\right)D_L\right].$$

(B.7)

The cost of serving the debt has systematically dropped. Again, D_B, and therefore the borrower's IRR, rises as D_T rises or D_L (that is, r_L) falls. How does the debt share affect the borrower's IRR?

$$\frac{\partial D_B}{\partial s} = \left[\left(D_L s - \frac{N_B}{I}\right)\frac{\partial p}{\partial s} + \left(\frac{D_T - \left(1 - p\right)D_L}{1 - s}\right)\right]\left(\frac{1}{1-s}\right).$$

(B.8)

Comparing this to Equation B.4, we see two important changes. First, when debt share has no effect on probability of default, the benefit from expanding debt share has risen, because the cost of debt to the borrower has systematically dropped at every level of debt share. In Figure B.1, the curve showing

$$\frac{D_T - D_L}{(1-s)_O}$$

effectively moves up to

$$\frac{D_T - (1-p)D_L}{(1-s)_G}.$$

Second, when increasing debt share increases the probability of default, this increase drops the effective cost of debt capital still further. Figure B.1 displays this effect as a downward movement in the

$$\left[\frac{N_B}{I}\left(\frac{\partial p}{\partial s}\right)\right]_O \quad \text{to} \quad \left[\left(\frac{N_B}{I} - D_L s\right)\frac{\partial p}{\partial s}\right]_G.$$

This curve can actually fall below the horizontal axis, as shown. As s and $\partial p/\partial s$ rise with rising debt share, however, it should become positive, making possible a well-defined optimum at $s_G > s_O$. In sum, the availability of a loan guarantee encourages a borrower to increase its debt share.

When m_{LO}, c_{LO}, and c_B are nonzero, the expressions get much more complicated, but the qualitative results are similar. Now

$$(1-s)ID_B = (1-p)\left[N_B - D_L\left(1 + m_{LG} + pc_{LG}\right)sI\right] + pD_B c_B sI \quad \text{and}$$

$$D_B = \frac{D_T}{1-s-pc_B s} - (1-p)D_L\left(1 + m_{LG} + pc_{LG}\right)\left(\frac{s}{1-s-pc_B s}\right). \tag{B.9}$$

Again, the definitions of annual payments relevant to the lender and borrower have become complex, but the structure of Equation B.9 is the same as that for Equation B.7. Again, not only D_T and p but also m_{LG}, c_{LG}, and c_B are functions of debt share. These additional factors tend to have smaller effects than the ones examined earlier. Again, including them leads to a lower optimal debt share simply because each of them involves costs associated with using debt financing, and each of these costs tends to rise as the debt share rises. Otherwise, including them does not change the qualitative findings presented here or the intuition underlying them.

Again, the difference between a situation without a loan guarantee and that with one occurs *entirely* because a higher debt share increases the likelihood of default and the cost of default is lower for the borrower with a loan guarantee than without. The government foots the bill when the borrower pursues a policy that increases the likelihood of default.

To see what it means for the government to foot the bill, compare the expressions in rows 9 and 12 to see that the borrower expects to pay (B) and what the government expects to pay (G) under a loan guarantee:

$$B = \left(1 - p\right)D_L\left(1 + m_{LG} + pc_{LG}\right)sI + D_B pc_B sI$$
$$G = pD_L\left(1 + m_{LG} + pc_{LG}\right)sI + \left(pD_G c_G + D_G m_{GG}\right)sI \tag{B.10}$$

Rearranging the terms in Equation B.10 yields the following:

$$G = \frac{pB}{1 - p} + p\left(c_G D_G - c_B D_B\right)sI + m_{GG}D_G sI. \tag{B.11}$$

In any year, the government expects to pay $p/(1 - p)$ times what the borrower pays, with two adjustments:

adjustment 1: $+ p(c_G D_G - c_B D_B)sI$
adjustment 2: $+ m_{GG}D_G sI$.

It may pay more or less than this depending on whether the government or borrower suffers more when a default actually occurs (adjustment 1). And it pays an additional amount to cover monitoring costs (adjustment 2). Unless the probability of default is very small, these two adjustments are likely to be small relative to the first term on the right in Equation B.11. The cash-flow model is not designed to track either of these adjustment terms. To a first order of magnitude, it is reasonable to expect that the government expects to pay about $p/(1 - p)$ times what the cash-flow model indicates that the borrower expects to pay in any year to service its debt. This is equivalent to saying that, to a first order of magnitude, the government simply expects whatever debt-related costs the borrower expects not to pay.

Table B.2 offers another way to see this point. It sets all costs of monitoring and default shown in Table B.1 to zero and then organizes the expressions from Table B.1 such that annual cash flows without a loan guarantee appear in the top half of the table and annual cash flows with a loan guarantee appear in the bottom. Columns show cash flows without and with default and then the expected cash flows. Rows show cash flows for the lender, borrower, and government and then a total of all three.

Focus on the last column, which displays expected values of cash flows. The lender expects the same income with and without a loan guarantee. With a loan guarantee, the borrower expects the government to pay a portion of the lender's income, effectively reducing the portion that the borrower must pay and so reducing the lender's cost of borrowing. This change is the principal effect of introducing a loan guarantee.

Table B.2
Central Effects of Introducing a Loan Guarantee

Perspective	No Default	Default	Expected
Probability	$1-p$	p	1
Without a loan guarantee			
Lender income	$\dfrac{D_L sl}{1-p}$	0	$D_L sl$
Borrower income	$\dfrac{N_B - D_L sl}{1-p}$	0	$(1-p)N_B - D_L sl$
Government payment	0	0	0
Total	N_B	0	$(1-p)N_B$
With a loan guarantee			
Lender income	$D_L sl$	$D_L sl$	$D_L sl$
Borrower income	$N_B - D_L sl$	0	$(1-p)(N_B - D_L sl)$
Government payment	0	$D_L sl$	$p*D_L sl$
Total	N_B	0	$(1-p)N_B$

References

Alic, John A., David C. Mowery, and Edward S. Rubin, *U.S. Technology and Innovation Policies: Lessons for Climate Change*, Arlington, Va.: Pew Center on Global Climate Change, November 2003.

Anderson, Brig. Gen. Frank J. Jr., *A Plan to Accelerate the Transition to Performance-Based Services: Report of the 912(c) Study Group for Review of the Acquisition Training, Processes, and Tools for Services Contracts*, Fort Belvoir, Va.: Fort Belvoir Defense Technical Information Center, AF903T1, June 25, 1999. As of June 27, 2008:
http://handle.dtic.mil/100.2/ADA395925

Arimura, Toshi H., Akira Hibiki, and Nick Johnstone, "An Empirical Study of Environmental R&D: What Encourages Facilities to Be Environmentally Innovative?" presentation to OECD Conference on Public Environmental Policy and the Private Firm, Washington, D.C., June 14–15, 2005. As of June 27, 2008:
http://www.oecd.org/dataoecd/14/43/35120613.pdf

Arrow, Kenneth J., and Robert C. Lind, "Uncertainty and the Evaluation of Public Investment Decisions," *American Economic Review*, Vol. 60, No. 3, June 1970, pp. 364–378.

Bartis, James T., Frank Camm, and David S. Ortiz, *Producing Liquid Fuels from Coal: Prospects and Policy Issues*, Santa Monica, Calif.: RAND Corporation, MG-754-AF/NETL, 2008.

Bartis, James T., Tom LaTourrette, Lloyd Dixon, D. J. Peterson, and Gary Cecchine, *Oil Shale Development in the United States: Prospects and Policy Issues*, Santa Monica, Calif.: RAND Corporation, MG-414-NETL, 2005. As of June 27, 2008:
http://www.rand.org/pubs/monographs/MG414/

Blyth, W., and M. Yang, *Impact of Climate Change Policy Uncertainty on Power Investment*, Paris: International Energy Agency, IEA/SLT(2006)11, 2006.

Bolton, Patrick, and Mathias Dewatripont, *Contract Theory*, Cambridge, Mass.: MIT Press, 2005.

Camm, Frank, Irv Blickstein, and Jose Venzor, *Recent Large Service Acquisitions in the Department of Defense: Lessons for the Office of the Secretary of Defense*, Santa Monica, Calif.: RAND Corporation, MG-107-OSD, 2004. As of June 30, 2008:
http://www.rand.org/pubs/monographs/MG107/

Dixit, Avinash, "Incentives and Organizations in the Public Sector: An Interpretative Review," *Journal of Human Resources*, Vol. 37, No. 4, Autumn 2002, pp. 696–727.

EIA—*see* Energy Information Administration.

Energy Information Administration, *Annual Energy Outlook 2007 with Projections for 2030*, Washington, D.C., DOE/EIA-0383(2007), February 2007. As of June 23, 2008:
http://www.eia.doe.gov/oiaf/archive/aeo07/

Farrar, Gary, "Processing: Nelson-Farrar Quarterly Costimating: Indexes for Selected Equipment Show Moderate Increase," *Oil and Gas Journal*, Vol. 105, No. 13, April 2, 2007, pp. 54–55.

Goldberg, Victor P., *Readings in the Economics of Contract Law*, Cambridge and New York: Cambridge University Press, 1989.

GSA, DoD, and NASA—*see* U.S. General Services Administration, U.S. Department of Defense, and National Aeronautics and Space Administration.

Gupta, Sujata, Dennis A. Tirpak, Nicholas Burger, Joyeeta Gupta, Niklas Höhne, Antonina Ivanova Boncheva, Gorashi Mohammed Kanoan, Charles Kolstad, Joseph A. Kruger, Axel Michaelowa, Shinya Murase, Jonathan Pershing, Tatsuyoshi Saijo, Agus Sari, Michel den Elzen, and Hongwei Yang, "Policies, Instruments and Co-Operative Arrangements," in Intergovernmental Panel on Climate Change, *Climate Change 2007: Mitigation of Climate Change: Contribution of Working Group III to the Fourth Assessment Report of the Intergovernmental Panel on Climate Change*, Cambridge and New York: Cambridge University Press, 2007, pp. 745–808. As of June 27, 2008:
http://www.ipcc.ch/pdf/assessment-report/ar4/wg3/ar4-wg3-chapter13.pdf

Hamilton, Kirsty, "The 'Finance-Policy' Gap: Policy Conditions for Attracting Investment," in *The Finance of Climate Change: A Guide for Governments, Corporations and Investors*, London: Risk Books, 2005.

Internal Revenue Service, *How to Depreciate Property: Section 179 Deductions, MACRS, Listed Property*, Washington, D.C.: Department of the Treasury, Internal Revenue Service, publication 946(2007), 2007. As of June 29, 2008:
http://purl.access.gpo.gov/GPO/LPS8194

IRS—*see* Internal Revenue Service.

Joskow, Paul L., "Contract Duration and Relationship-Specific Investments: Empirical Evidence from Coal Markets," *American Economic Review*, Vol. 77, No. 1, March 1987, pp. 168–185.

Kretzschmar, Gavin Lee, and Axel Kirchner, "Commodity Price Shocks and Economic State Variables: Empirical Insights into Asset Valuation and Risk in the Oil and Gas Sector," *Journal of Banking and Finance* Conference, Commodities and Finance Centre, University of London, January 2007.

Laffont, Jean-Jacques, and Jean Tirole, *A Theory of Incentives in Procurement and Regulation*, Cambridge, Mass.: MIT Press, 1993.

Lempert, Robert J., Steven W. Popper, and Steven C. Bankes, *Shaping the Next One Hundred Years: New Methods for Quantitative, Long-Term Policy Analysis*, Santa Monica, Calif.: RAND Corporation, MR-1626-RPC, 2003. As of June 29, 2008:
http://www.rand.org/pubs/monograph_reports/MR1626/

Lewis, Frank, and Mary MacKinnon, "Government Loan Guarantees and the Failure of the Canadian Northern Railway," *Journal of Economic History*, Vol. 47, No. 1, March 1987, pp. 175–196.

Masten, Scott E., "Contractual Choice," in Boudewijn Bouckaert and Gerrit De Geest, eds., *Encyclopedia of Law and Economics*, Vol. III: *The Regulation of Contracts*, Cheltenham, Edward Elgar, 2000. As of June 25, 2008:
http://encyclo.findlaw.com/4100book.pdf

Merrow, Edward W., Kenneth Phillips, and Christopher W. Myers, *Understanding Cost Growth and Performance Shortfalls in Pioneer Process Plants*, Santa Monica, Calif.: RAND Corporation, R-2569-DOE, 1981. As of June 27, 2008:
http://www.rand.org/pubs/reports/R2569/

Metcalf, Gilbert E., *Federal Tax Policy Towards Energy*, Cambridge, Mass.: National Bureau of Economic Research, working paper 12568, October 2006. As of June 27, 2008:
http://www.nber.org/papers/w12568

National Energy Technology Laboratory, *Baseline Technical and Economic Assessment of a Commercial Scale Fischer-Tropsch Liquids Facility*, Washington, D.C.: U.S. Department of Energy, DOE/NETL-2007/1260, April 9, 2007. As of June 23, 2008:
http://204.154.137.14/energy-analyses/pubs/Baseline%20Technical%20and%20Economic%20Assessment%20of%20a%20Commercial%20S.pdf

NETL—*see* National Energy Technology Laboratory.

Office of Management and Budget, *Guidelines and Discount Rates for Benefit-Cost Analysis of Federal Programs*, Washington, D.C.: Executive Office of the President, Office of Management and Budget, OMB Circular A-94 (rev.), October 29, 1992.

OMB—*see* Office of Management and Budget.

Peterson, D. J., Tom LaTourrette, and James T. Bartis, *New Forces at Work in Mining: Industry Views of Critical Technologies*, Santa Monica, Calif.: RAND Corporation, MR-1324-OSTP, 2001. As of June 27, 2008: http://www.rand.org/pubs/monograph_reports/MR1324/

Reedman, Luke, Paul Graham, and Peter Coombes, "Using a Real-Options Approach to Model Technology Adoption Under Carbon Price Uncertainty: An Application to the Australian Electricity Generation Sector," *Economic Record*, Vol. 82, No. S1, September 2006, pp. S64–S73.

Rubin, Paul H., *Managing Business Transactions: Controlling the Cost of Coordinating, Communicating, and Decision Making*, New York: Free Press, 1990.

Salanié, Bernard, *The Economics of Contracts: A Primer*, 2nd ed., Cambridge, Mass.: MIT Press, 2005.

Southern States Energy Board, *American Energy Security: Building a Bridge to Energy Independence and to a Sustainable Energy Future*, Norcross, Ga., July 2006. As of June 23, 2008: http://www.americanenergysecurity.org/studyrelease.html

SSEB—*see* Southern States Energy Board.

U.S. General Services Administration, U.S. Department of Defense, and National Aeronautics and Space Administration, *Federal Acquisition Regulation*, Washington, D.C.: U.S. General Services Administration, effective June 12, 2008. As of June 30, 2008: http://acquisition.gov/far/

WTRG Economics, "Oil Price History and Analysis (Updating)," undated Web page. As of March 18, 2008: http://www.wtrg.com/prices.htm